Advances in
Nuclear Science
and Technology

VOLUME 17

Simulators for Nuclear Power

Advances in Nuclear Science and Technology

Series Editors

Jeffery Lewins
Cambridge University, Cambridge, England

Martin Becker
Rensselaer Polytechnic Institute, Troy, New York

Editorial Board

Eugene P. Wigner, *Honorary President*
R. W. Albrecht
F. R. Farmer
Ernest J. Henley
Norman Hilberry
John D. McKean
ʻK. Oshima
A. Sesonske
H. B. Smets
Karl Wirtz
C. P. L. Zaleski

Advances in Nuclear Science and Technology

VOLUME 17

Simulators for Nuclear Power

Edited by

Jeffery Lewins
Fellow of Magdalene College
Cambridge University
Cambridge, England

and

Martin Becker
Rensselaer Polytechnic Instiute
Troy, New York

PLENUM PRESS · NEW YORK AND LONDON

Library of Congress Cataloging in Publication Data

Main entry under title:

Simulators for nuclear power.

(Advances in nuclear science and technology; v. 17)
Includes bibliographies and index.
Contents: A methodology for the design of plant analysers/T. H. E. Chambers
and M. J. Whitmarsh-Everiss—Models and simulation in nuclear power station
design and operation/M. W. Jervis—Psychological aspects of simulator design and
use/R. B. Stammers—[etc.]
1. Nuclear power plants—Simulation meethods. I. Lewins, Jeffery. II. Becker,
Martin. III. Series.
TK9001.A3 vol. 17 [TK9153]   621.48s   85-31170
ISBN-13: 978-1-4612-9285-2   e-ISBN-13: 978-1-4613-2183-5
DOI: 10.1007/978-1-4613-2183-5

© 1986 Plenum Press, New York
Softcover reprint of the hardcover 1st edition 1986
A Division of Plenum Publishing Corporation
233 Spring Street, New York, N.Y. 10013

PREFACE

      This volume represents the second of our occasional departures from the format of an annual review series, being devoted to one coherent topic. We have the pleasure therefore in presenting a concerted sequence of articles on the use of Simulators for Nuclear Power.

      An essential attribute of a quantified engineer in any discipline is to be able to model and predict, i.e. to analyze, the behaviour of the subject under scrutiny. Simulation goes, one would argue, a step further. The engineer providing a simulator takes a broader view of the system studied and makes the analysis available to a wider audience. Hence simulation may have a part to play in design but also in operation, in accident studies and also in training. It leads to synthesis as well as analysis.

      There is no doubt that the massive scale and the economic investment implied in nuclear power programmes demands an increased infra-structure in licensing and training as well as in design and operation. The simulator is a cheap alternative - admittedly cheap only in relative terms - but also perhaps an essential method of providing realistic experience with negligible or at least small risk.

      Nuclear power therefore has led to a wide range of simulators. At the same time we would not overlook the substantial role played by simulators in say the aero-industry; indeed the ergonomic and psychological studies associated with that industry hold many lessons.

      In this volume we have sought to provide a coherent account of the approach necessary to commission simulators on the scale required in civil nuclear power applications. To do this we have drawn on a number of authors who have been associated with the UK nuclear power program, in the Central Electricity Generating Board responsible for the reactors

themselves, with their associates in the South of Scotland
Electricity Generating Board and with the associated contractors and university research staff. We are particularly
grateful to the CEGB that we can publish an account of the
philosophy and methodology of the development of simulators
in one of the world's largest interconnected utilities. They
have enabled all our authors to elucidate the systems thinking
necessary to implement simulation on a scale that encompasses
tens of nuclear plants and a fifty years planning horizon.

Many established nuclear utilities are now coming to
grips with the transition to a 'steady state' in which the
operation, refuelling and ultimate decommissioning of nuclear
plant looms as large in corporate planning as design and
construction. It is easy to forget that over a forty-year
life, a nuclear reactor will have as much in money terms
flowing through it as fuel as it did initially as construction
capital. Correspondingly, the training of operators - to be
renewed and replaced over a forty year life - is as significant as initial design. Aids for prediction in normal and
emergency circumstances, which must run faster than real
time, are as important as design simulations.

The large scope and far planning horizon of a nuclear
power plant calls for suitable management methods to ensure
satisfactory performance from simulators which themselves
represent an investment of millions of dollars. In general
terms, the UK, like many other developed countries using
'high tech', is better at the initial entrepreneurial development and not so good in the management aspect of seeing that
new engineers come forward to complete and maintain the high
tech systems. Thus a methodology for high tech simulation,
a Quality Assurance programme and a management approach are
essential ingredients to success.

It seemed to your editors that an account of such a
philosophy would be valuable and our first two articles, by
CEGB authors, are directed to that understanding of how a
large organization might tackle the management of simulation
problem.

The third article turns to the psychological problems of
simulation in terms of the viewer and user rather than the
organization. Before committing large resources, it is as
well for the proprietor to establish what it is he wants and
why a simulator might provide it.

Our fourth article is again authored from within the CEGB but addresses in more explicit terms how the computation for simulation of nuclear plant can be carried out in relation to the rapdily changing availability of computing facilities. A particular feature of CEGB experience is the driving of training simulators from mainframes more than 200 km away

New developments will come in simulation as well as in computing power. But the need for real time or even better computation with the detail required in such complex plants, for training and emergency situations, puts a premium on new thinking. One current and promising development is in the use of parallel processing; our second contribution from the universities addresses this area with specific reference to two-phase flow calculations. The preliminary results reported are indeed encouraging.

Parallel processing is in part here with us in the use of multiple microprocessors. The final article of the volume puts both present and future practice together with an account of the construction of the SSEBs simulators with their parallel processing architecture and the possibilities of future development.

To obtain the coherent picture of nuclear simulators provided here we have drawn our authors from the UK with its experience of gas-cooled reactors. But the more important features are the scale of the operations involved rather than the particular reactor design and it is this scale of enterprise with its concomitant demands on technical management that we have sought to put before our readers.

Jeffery Lewins and
Martin Becker

August, 1985

CONTENTS

A Methodology for the Design of Plant Analysers
    T. H. E. Chambers and M. J. Whitmarsh-Everiss

Models and Simulation in Nuclear Power Station Design and
Operation
    M. W. Jervis

Psychological Aspects of Simulator Design and Use
       R. B. Stammers

The Development of Full-Scope AGR Training Simulators
within the CEGB
        G. C. Budd

Parallel Processing for Nuclear Safety Simulation
        A. Y. Allidina, M. G. Singh and B. Daniels

CONTENTS OF EARLIER VOLUMES

Volumes 1-9 of *Advances in Nuclear Science and Technology*
were published by Academic Press, New York.

SI UNITS

It is twenty five years from the adoption of the internationally agreed system of units, the SI Units or Le Système International d'Unités. Our series uses SI units to facilitate the international exchange of information and learning we exist to promote. A few words on the system may be of interest to some of our readers and, of course, to our potential authors.

The system was adopted by the 1960 meeting of the General Conference of Weights and Measures. The international Standards Organisation (ISO) publishes the fundamental documents detailing the system in a series of reports under the number R 31. Part 0 deals with general principles, Part I with the fundamental units and so on. Of particular interest might be: Part IV, heat; Part IX, atomic and nuclear physics, Part X, nuclear reactions. Of course changes of detail authorised at subsequent meetings have occurred. It is usually best to obtain up to date information from the relevant national standards organisation.

Some particular changes to note include the following. The symbol for the litre may now be taken as either L or l We prefer the capital L since the lower case l may well be confused, especially in typescript, with 1 although we note the breach of the convention that only eponymous symbols should be capitalised. *En passant*, we note that the correct abbreviation for second is s and not sec.

The SI units for health physics with special names have been agreed more recently. The significant units are:

Activity: becquerel (Bq) 1 disintegration per second. This has replaced the obsolescent unit of the curie (Ci) where 1 Ci = 37 GBq.

Absorbed Dose: gray (Gy) 1 joule per kilogram. This has replaced the obsolescent unit of the rad (100 rad = 1 Gy).

Equivalent Absorbed Dose: sievert (Sv) also 1 joule per kilogram. This has replaced the obsolescent unit of the rem (100 rem = 1 Sv).

In some applications, large quantities of becquerels etc. are involved. Recently the following size modifiers have been authorised in the SI canon and would be useful in this context:

E (exa) 10^{18} : P (peta) 10^{15}.

Nuclear technology is bound to the wheel of its own history. It will take some time for the older units to be phased out and for new dogs to come with new tricks. The universities have a particular responsibility here to be didactic. It will not of course be until the old units have been replaced in national legislation that they can be said to be obsolete rather than obsolescent. The charm of the barn, 10^{-28} m^2, is likely to stay with us for a long time despite its deviation from SI norms.

There are other individual areas where we must suppose practice has not made for perfection. It is of course time for British Units to be dropped. One can see that if a reactor has been designed with a six-inch diameter pitch or coolant pipe etc., then it is quixotic to solemnly translate this design into meters but for international understanding perhaps the original design value should be followed by its SI equivalent in what we hope will be limited cases. There is really no case for the retention of British Thermal Units or the Fahrenheit/Rankine temperature scale. Rather we prefer the kelvin unit and speak of a temperature of (say) 300 K.

It is of course customary to use an offset or celsius scale analogous to a reference to gauge pressure or (absolute) pressure. The symbol for the celsius temperature is $^{\circ}C$ as in say 100 $^{\circ}C$. The unit of interval on the celsius scale however is of course the kelvin (K). The 'degree' sign is only necessary in conjunction with the abbreviated symbol of which it is part to distinguish $^{\circ}C$ from C or coulomb; there is no place for the word 'degree'.

The SI system calls for the nature of what is quantified to be written *before* the quantity, followed by the SI units

in suitable multiples. It would be good practice then to
write a neutron flux as

$$\text{neutron flux } 10^{17} \text{ m}^{-2} \text{ s}^{-1}$$

and *not* as neutron/m^{2}s. One notes that symbols are not
pluralised. It would follow (but is not yet recognised
practice) that a gauge pressure might be written

$$\text{gauge pressure 10 Pa}$$

and *not* 10 Pa gauge while a celsius temperature might by
analogy be written as

$$\text{celsius temperature } -10 \text{ K}$$

and *not* -10 $^{\circ}$C.

   This principle of placing only the SI units to the
right of the quantity raises a difficulty in referring to the
power output of a nuclear plant. Here one desires to
distinguish between the thermal and the mechanical/electrical
power delivered. Because we lack the established words to
distinguish these concepts we are driven to the neologisms of
MW(t) and MW(e) in the units against good SI practice. If
it is energy to be distinguished in this way we can suggest
the European usage of exergy to refer to the equivalent as
mechanical and hence electrical work. We could then write in
terms of an energy of so many joules for the (total) or an
exergy of so many joules for the mechanical element. We
tentatively suggest the term *rexergy* (rate of exergy) for the
corresponding rate of transmission of energy in its equivalent mechanical form so that we might refer to (say)

$$\text{reactor power 3000 MW and reactor rexergy 1000 MW.}$$

This 'kite' is flown with the intention of stimulating
discussion and perhaps better suggestions.

   National practice has differed in the writing of large
numbers with some using a '.' and some a ',' for the decimal
point. Thus it may avoid confusion with a decimal point to
use blank spaces only to separate trios of figures as in
11 000.001 01. Tables should be prepared accordingly.

   Economy in the writing of numbers is sometimes available.

For example, it seems pointless to write typical values of
fuel burn-up as 30 000 MWd/tonne when the same quantity might
be written 30 MWd/kg. Here we accept that Wd is a more
meaningful but less basic unit to be employed rather than
(24 x 3600) J. Strictly there should be space between units
and we should write W d *not* Wd.

Some aspects of the technical writers craft are not yet
covered in SI but various national handbooks amplify good
practice. We think particularly of the Royal Society of
London's useful booklet "Quantities, Units and Symbols"
(1975). We adopt their advice over the labelling of the axes
of graphs. If an axis is shown as say pressure divided by its
unit, P/Pa, then the numbers used along the axis are properly
non-dimensional.

Much can be done at the editorial stage to provide for
a consistent usage of SI units but it is even more economical
if our authors would cooperate in making their work as widely
available as they deserve.

A METHODOLOGY FOR THE DESIGN OF PLANT ANALYSERS

T.H.E. Chambers and M.J. Whitmarsh-Everiss

Central Electricity Generating Board, U.K

I. INTRODUCTION

The design and provision of plant analysers or engineering simulators, as they are variously known, is seen as
being no more than the logical conclusion to the evolution
of the design methodology embracing control system design,
plant operational dynamics and fault analysis. Over the
past twenty-five years this methodology has been evolving to
meet the requirements of an advancing technology, but
heavily constrained in its development by evolving numerical
methods and computing architectures. For this reason it is
instructive to consider its historical development. It has
of course occurred internationally but for the purposes of
illustration, only the development within the Central
Electricity Generating Board (CEGB) will be considered,
although it should be understood that this has been conditioned particularly by developments in the USA and the British
Nuclear Power Industry.

As this historical development unfolds the logic behind
this converging methodology will become clear, as will the
universality of its approach. By the latter is meant its
ability to handle any type of plant, whether conventional
or nuclear and for whatever purpose e.g. for design or
simulators for operator training.

It should be recognised that the CEGB provides a unique
environment in which design and training methodologies can
develop, flourish and be consolidated through considerable
investment in applied research and plant operational feedback.

1

We will now consider the relevant aspects of the CEGB operation in order to establish the environment in which the historical development of their simulation methodology has taken place and hence underwrite its validity.

II. CEGB OPERATIONS

The Central Electricity Generating Board is the sole authority responsible for the planning, production and distribution of electrical supplies for the whole of England and Wales. It is also charged with the responsibility for the design, procurement, construction, commissioning and operation of the generating stations required to meet consumer demand within a framework laid down by Statute of Parliament.

The generating plant extant on the Board's grid system comprises a mixture of nuclear, fossil-fired, gas turbine, hydro and pumped storage. The contracts for conventional plant have been let to the major component manufacturers for civil works, boiler, turbine, etc., the Board retaining responsibility for Project Management and integration of the various sub-systems to itself. Historically, nuclear plants have been let under a turnkey contract to nuclear consortia, acting as Architect Engineers and originally consisting of partners made up from the manufacturers. The most recent nuclear station, Heysham II has been organised more like the conventional stations - the Board maintaining responsibility for the turbo-generator, civil works, etc. but purchasing a nuclear steam supply system from the one remaining Consortium. For various reasons the Board has been very closely involved in the design process for nuclear plant. The Board is required to justify the safety case for each station to the Nuclear Installations Inspectorate and consequently is concerned with all apsects leading to the request for a Site Licence to operate. This has far reaching implications extending to involvement with plant modelling, fundamental research into materials properties, specific component testing and the commissioning of experimental work when unforeseen problems have arisen during construction.

The Board's perspective of the plant is somewhat different to the manufacturers' in that the capital required for the plant's construction has to be justified as an investment and hence the return over the design life must be ensured, as must its operability and maintainability. As a consequence of these considerations the Board has considerable independent expertise.

Work in the area of plant modelling originated from the separate issues of safety on the one hand and control and operational dynamics on the other. These two areas are now seen to be converging. The principles upon which the models were built has always been the same, although they differed in the way the data was handled. Total plant models of varying degrees of asymmetry have been developed for all the Advanced Gas Cooled Reactors; generally they comprise an axially distributed single channel reactor model, coupled to one or more axially distributed boilers, turbines and associated feed pumps, feed trains, gas circulators, control system and protection systems.

For deeper study of particular components more detailed separate models are available giving better definition of axial, x-y, radial/azimuthal or other characteristics.

The Board's laboratories are involved in a continuing programme of research into the material properties, reactor physics and gas dynamics, chemical and metallurgical characteristics of the various solids, fluids and gases currently being utilised in the nuclear stations. In many instances the work is directed at a particular geometry/fluid, for example the heat transfer characteristics of a two-phase water/steam mixture flowing in a serpentine geometry, or the mixing characteristics of carbon dioxide flowing over banks of serpentine tubes. The results of this type of rig work are used subsequently in conjunction with the models mentioned earlier in a predictive capacity for design studies. For components which it is not possible to emulate in experimental rigs, works test programmes are formulated with the manufacturers with a view to establishing the characteristics in a regime as near as possible to that which would pertain in operation on the plant, or alternatively utilising the concept of dimensional similarity. The design process is essentially iterative particularly when each plant is significantly different from another. In order to maximise the confidence in model predictions, test programmes are set up to be implemented during plant commissioning. The procedure is collaborative with a joint team drawn from the Board, Consortium and manufacturers. Model predictions can then be compared against the plant tests and then modified and improved as necessary.

When commissioning is completed, the plant is administratively handed over to the geographical region in which

TABLE 1

SCHEDULE OF U.K. POWER REACTORS, OPERATING,
COMMISSIONING AND UNDER CONSTRUCTION,TYPE,
OUTPUT AND KEY PROGRAMME DATES

A.   United Kingdom Atomic Energy Authority

Station	Type	Power Output Gross MW	Number of Reactors	Raise Power Date
Calder Hall	Magnox	240	4	1956-59
Chapelcross	Magnox	240	4	1958-60
Windscale	AGR	32	1	1963
SGHWR Winfrith	SGHWR	100	1	1968
Dounreay	DFR	15	1	1963
Dounreay	PFR	270	1	1976

B.   Central Electricity Generating Board

Station	Type	Power Output Gross MW	Number of Reactors	Raise Power Date
Berkeley	Magnox	276	2	1962
Bradwell	Magnox	300	2	1962
Hinkley Point'A'	Magnox	500	2	1965
Trawsfynydd	Magnox	500	2	1965
Dungeness 'A'	Magnox	550	2	1965
Sizewell 'A'	Magnox	500	2	1966
Oldbury	Magnox	416	2	1967
Wylfa	Magnox	840	2	1971
Hinkley Point'B'	AGR	1040	2	1976
Dungeness 'B'	AGR	1200	) 2Commis-1983	
Hartlepool	AGR	1332	) 2sioning1983	
Heysham I	AGR	1332	) 2	1983
Heysham II	AGR	1330	2	Under construction

C.   South of Scotland Electricity Board

Station	Type	Power Output Gross MW	Number of Reactors	Raise Power Date
Hunterston 'A'	Magnox	334	2	1964
Hunterston 'B'	AGR	1040	2	1976-7
Torness	AGR	1320	2	Under construction

D. Schedule of Modern Coal and Oil Fired Plant on the
    CEGB System

Station	Number of Sets	Total power Output MWSO	Fuel	Date of Synchronis-ation
Aberthaw 'B'	3 x 500 MW	1330	Coal	1968-71
Cottam	4 x 500 MW	1840	Coal	1968-70
Didcot	4 x 500 MW	1820	Coal	1970-74
Drax I	3 x 660 MW	1875	Coal	1973-74
Drax Completion	3 x 660 MW	Data not yet available	Coal	1983-
Eggborough	4 x 500 MW	1720	Coal	1967-69
Fawley	4 x 500 MW	1932	Oil	1969-71
Ferrybridge 'C'	4 x 500 MW	1932	Coal	1966-67
Fiddlers Ferry	4 x 500 MW	1880	Coal	1970-73
Grain	4 x 660 MW	2640	Oil	1979
Ince 'B'	2 x 500 MW	960	Oil	1980-81
Ironbridge 'B'	2 x 500 MW	920	Coal	1969-70
Kingsnorth	4 x 500 MW	1920	Coal/ Oil	1970-73
Littlebrook 'D'	3 x 660 MW	1980	Oil	1981
Pembroke	4 x 500 MW	1900	Oil	1970-72
Ratcliffe	4 x 500 MW	1932	Coal	1967-70
Rugeley 'B'	2 x 500 MW	920	Coal	1967-70
West Burton	4 x 500 MW	1840	Coal	1966-68

it lies. The Region is then responsible for operation on a
daily basis. To assist in this role the appropriate Regional
Scientific Services Division or Engineering Department can
advise on or undertake specific testing, the provision of
instrumentation, or provide models to cater for problems as
they arise. The Board is thus seen to be actively involved
in the design, commissioning and operation of its stations
and so is in a position to gain experience in, and hence
obtain feedback, from all these phases. Future work, relating to any type of system, will be organised along similar
lines.

In the above statement, the AGR was used by way of
example only. The Board has, of course, considerable
experience in the design and operation of Magnox plant and
coal and oil fired conventional plant. It has also been
involved with the UKAEA in design and operational considerations in respect of the Calder Hall and Chapelcross Magnox
plants, the HTR, SGHWR, DFR and PFR. The extent of this
involvement is illustrated in Tables 1A-D.

In addition detailed design studies have been mounted
for proposed HTR, SGHWR, CFR and PWR plant. The PWR is
currently the subject of a Public Enquiry seeking a site
licence for Sizewell 'B'.

III. THE HISTORICAL DEVELOPMENT OF DESIGN AND PLANT
SIMULATION METHODOLOGY WITHIN THE CEGB

In 1959 the only computer available in the Board for
design analysis was the digital English Electric Deuce.
From this was obtained reactor axial steady states of trivial
complexity. No dynamic analysis was attempted. Early in
1962 the Bradwell and Berkeley Magnox plants were at an
advanced stage of construction and design problems and licensing responsibilities justified the acquisition of the EAL
PACE 231R analogue computer having 200 amplifiers. This was
used extensively for the Magnox reactor safety case using
one-dimensional axially distributed reactor models for
steady state and dynamic analysis. Subsequent modelling
included of radial/azimuthal modal analysis and boiler steady
state calculations. The latter calculation used the ADIOS
logic facility to iterate to, and satisfy the boiler boundary
conditions. The logical extension of this type of calculation
was the continuous space, discrete time, dynamic solution of
the equations and this was attempted on an early hybrid

computer (1965) but without success. The 231R facility was used extensively until 1969. Modelling was developed for one dimensional axially distributed AGR reactors, very simple total plant models for both conventional and nuclear plant, and for grid system and electrical system dynamic analysis.

Similar facilities were available in the Atomic Energy Authority and in the Consortia and one development in particular is worthy of note. In 1957 the Consortium, now known as NNC, constructed at Whetstone an analogue computer known as PLUTO having 115 amplifiers. This was followed in 1959 by MARS, having 210 amplifiers. Both machines were phased out in 1959/1960 and replaced by SATURN having 6 consoles each having 252 amplifiers, 600 potentiometers and 76 non-linear devices. Over the years these consoles were trunked together to yield a 1500 amplifier facility which was subsequently supplemented in 1972 with an EAI 8812 analogue computer. It is believed that SATURN was the largest analogue computer ever built. Excellent service was given by this machine until 1978 when it was scrapped. These various analogues were used by the Consortium sited at Whetstone and by the Board for numerous control system design, plant operational and fault analyses. Detailed analysis was carried out for Hinkley Point 'A', Sizewell 'A', Wylfa, Hartlepool and Heysham I, whilst analysis for unsuccessful tender submissions was effected for Trawsfynydd, Dungeness 'A', Oldbury and Dungeness 'B'. In addition studies were made of HRT, SGHWR, PWR and CFR designs.

The Board continued to take an interest in the development of hybrid computer techniques and in 1966 a study of the Dungeness 'B' plant was carried out on the EAI 8800 hybrid computer at the EAI mathematical laboratory at Princeton USA. This study was successful and led to the Board installing the EAI 8800 system at its Computing Bureau in London in 1969. Before considering this system further it is necessary to consider the parallel digital computer and simulation language development.

In 1962 the Board installed an IBM 7090 digital computer. The Atomic Energy Authority had developed a 3-dimensional reactor dynamic model for Magnox plant in polar co-ordinates and this was run on the IBM 7090 for operational transient and fault analysis for both the Bradwell and Berkeley plants.

At this time there was a growing interest in the digital emulation of analogue models in codes such as DAS and MIDAS.

These were acquired by the Board but found to be too restrictive for general use. In 1964 the IBM 7090 system was updated
to the 7094 and in the same year the digital simulation language DSL90 was acquired from IBM and used for the total plant
modelling of Dungeness 'B'. This was successful, but very
expensive, as the code ran 30 times slower than real time
using the 5th order Milne integration algorithm. In 1966
the simulation language CSMP was acquired from IBM. This was
in use until 1970 when it was replaced by the Board's simulation language PMSP. In the meantime the IBM 7094 was
replaced by the IBM 360 system (1967), the word length decreasing from a 36 bit to a 32 bit word. This gave rise to
considerable difficulty which was overcome by the universal
application of double precision.

Experience with the PACE231R and SATURN analogue
computers suggested that the hybrid computer should be
provided with software for automatically setting up the
various models. This was done and equivalent models were
written in PMSP which yielded a static check, the initial
conditions for the integrators, and a dynamic check against
a standard run. Hybrid computer models were developed for
the Dungeness 'B', Hinkley Point 'B', Hartlepool and
Heysham I AGR plants. One-dimensional reactor models, dual
boilers and total plant models were developed. Non-linear
hill climbing techniques, SIMPLEX and COMPLEX were used for
automatic control system design and model fitting.

The hybrid computer was in fact a plant analyser. The
automatic setting up and static check software enabled a
model to be set up in 0.5-1 hr. The machine was normally
operated at 10 times faster than real time i.e. a beta of
0.1. Dependent upon the type of analysis some 200 runs a
day could be obtained. The output was handled through 24
channels of analogue recording, 4 x-y plotters and a high
speed digital printer. SATURN was run in a similar manner
except that a particular model had to be set up manually.
The setting up time was long (months) and hence all the
required information had to be obtained prior to breakdown
and re-application of the machine. Both machines were used
for control system design, operational studies and fault
analysis and allowed of extensive sensitivity surveys.
Station Operating Instructions and fault recovery procedures
were studied, and checked out in relation to ergonomic
considerations on desk and panel mockups. Neither machine
was ever used for advice to operating plant in the context

of operational problems, although they were used for the rapid resolution of problems during commissioning. The hybrid computer and the plant analyser have a problem in common which has not been solved. The problem arises from the enormous amount of data which is presented to the analyst which he cannot absorb in a timescale commensurate with a calculation which is being effected at a beta of 0.1. In the context of the plant analyser it is vitally important that this problem is resolved. This is particularly the case if the plant analyser is to be used in a role in which the analyst is to give on-line advice to plant operators on fault recovery procedures.

The hybrid computer could have been trunked up to 6 analogue consoles but it was realised that this technology would not advance and that it was preferable to create a wholly digital environment which would not restrict model size. The hybrid computer was therefore phased out in 1975 and SATURN was withdrawn in 1978. The replacement digital environment comprised the interactive editing language TSOeasy, the simulation language PMSP with fast integration algorithms, and the output grahics package VISION. No dedicated main frame computer was used, code being executed in batch. However, the main frames were continually updated as the machine architectures evolved, see Table 2, the speed increasing by a factor of 1.5-2.0 each time the machine was updated. This increase in speed and the development of fast integration algorithms has, in general, kept pace with increases in model size but the ability to run in real time and faster has never been recovered.

It is Board policy that all software must be upward compatible in the available main frame architectures. Whilst this conserves the enormous manpower investment in coding it militates against very fast solution times, i.e. beta less than 0.1, even though the Board expects to take advantage of the available parallel and pipeline machine architectures.

If very fast solutions of large codes conserving the design Quality Assurance route are required in the immediate future specialised computer architectures, languages and solution methods will have to be used. However, such developments are seen by the Board as having only short term value and the Board confidently expects that the evolution-ary development of models can be handled by developments in main frame computer design.

TABLE 2

CHRONOLOGICAL RELATIONSHIP BETWEEN MAIN FRAME DIGITAL, ANALOGUE AND HYBRID
COMPUTERS AVAILABLE TO THE CEGB FOR CONTROL SYSTEM DESIGN,
OPERATIONAL AND DYNAMIC ANALYSIS

Date	Main Frame Digital Computers	Analogue/ Hybrid Computers	Simulation Languages
1957			
1959	E.E. Deuce	PLUTO(NNC), analogue	
		MARS(NNC), analogue	
1960	IBM 709	PLUTO scrapped	
		SATURN(NNC), analogue	
		MARS scrapped	
9/1961	IBM 1401	EAL PACE 2 x 231R	
1/1962	IBM 7090	with ADIOS, analogue	
	(replacing 709)		
1963	ICL 1500		
1964	IBM 7094		DAS/MIDAS
	(replacing 7090)		
1965			
3/1966	IBM 360/30	Hybrid study	DSL/90 (IBM)
5/1966	IBM 360/30	at EAI Princeton	
8/1966	IBM 360/75	USA	CSMP (IBM)
	(1401 & 1500 removed)		

← 36 bit to 32 bit word

Double Precision

PMSP (CEGB)

Model complexity increasing →

Development of fast integration algorithms

EAL 8800 hybrid computer
EAL PACE analogue withdrawn
EAI 8812 (NNC)

EAL 8800 Hybrid Computer withdrawn

Saturn Analogue (NNC) scrapped

Machine speed increasing by factor of 1.5/annum →

1/1967	IBM 360/50 (replacing 360/30)
1/1968	7094 removed
7/1968	IBM 360/75
7/1968	IBM 360/50
12/1968	360/75 & 50 removed
1969	
1970	IBM 360/85 (replacing 360/75 & 50)
1972	IBM 370/165 (replacing 360/85)
7/1972	IBM 370/155 (replacing 360/155)
9/1974	IBM 370/158
1975	IBM 370/168 (replacing 360/165)
12/1976	IBM 370/168 (replacing 370/158)
1978	IBM 370/3032 (4M)
1979	IBM 370/3032 upgraded (6M)
1980	IBM 370/3032 upgraded (7M), AMDAHL V7 (8M)
1981	replacing 370/168, AMDAHL V7 upgraded (12M), AMDAHL V8 (16M)
1982	replacing V7, IBM 4341 (8M), IBM 3081 (24M)

IV. THE DESIGN AND PROVISION OF
PLANT TRAINING SIMULATORS

The CEGB has been involved in training simulator procurement and design since 1958 when a specification for a
generic Magnox recirculation drum boiler analogue simulator
was prepared, subsequently built by Elliots and commissioned
in 1959. This was supplemented in 1966 by an axially
distributed analogue reactor model engineered by Rediffusion.
This model was capable of being reconfigured to yield a
modal analysis model comprising 5 radial/azimuthal modes.
The modelling was based on the Pace 231R models then being
used for design purposes.

Subsequently the Education and Training Branch in the
North Eastern Region at Leeds designed a generic 500 MW coal/
oil fired analogue simulator. This was constructed in-house
by their Simulator Section and commissioned in 1971. This
was followed by a 660 MW coal or oil fired analogue/digital
simulator, again built in-house and commissioned in 1975.
The success of these two simulators led to a number of orders
for overseas plant namely for Escom in South Africa, 1975;
the Northern Ireland Electricity Service, 1978; the Bahrain
State Electricity Directorate, 1979 (this included a model of
a desalination plant); the State Electricity Commission,
Victoria, Australia, 1980; and Huntley Power Station, New
Zealand, 1981.

A new generation of digital simulators was produced
through the collaboration of the Leeds Simulator Section and
the Plant Kinetics Group at GDCD, Barnwood. This resulted
in the design of a coal/oil fired simulator for the Castle
Peak 'A' Power Station constructed for the China Light and
Power Company. The success of this simulator led to an
order for a training simulator for the Castle Peak 'B' plant.
This was completed in November 1984 and is thought to be the
most sophisticated conventional plant simulator ever built.

In addition to the above, the Leeds Simulator Section
have commissioned a 500 MW coal fired digital simulator for
in-house training purposes, whilst the South East Region
in collaboration with the CEGB Computing Centre have recently
commissioned an oil/coal fired digital simulator for
Littlebrook 'D'. These various digital simulators use two or
more SEL machines.

TABLE 3

PLANT TRAINING SIMULATORS DESIGNED FOR
IN-HOUSE USE OR UNDER CONTRACT BY THE CEGB

1.   Conventional Plant Simulators for in-house training

 (a) North East Region of CEGB Mark 1 training simulator.
   Generic 500 MW coal/oil fired analogue simulator.
   Commissioned in 1971. Engineered by NER CEGB.

 (b) North East Region of CEGB Mark II training simulator. Generic 660 MW coal fired or oil fired
   analogue/digital simulator. Commissioned in 1975.
   Engineered by NER CEGB.

 (c) North Eastern Region of CEGB Mark III training
   simulator. Generic 500 MW coal fired digital simulator. Commissioned - 1983/84. Engineered by
   NER CEGB.

 (d) South East Region of CEGB Littlebrook 'D' training
   simulator. 660 MW oil fired digital simulator.
   Commissioned 1983. Engineered by SER in collaboration with CEGB Computing Department and GDCD.

2.   Plant Simulators provided under contract

 (a) 500 MW unit coal fired analogue simulator for ESCOM,
   South Africa. Commissioned in 1975. Engineered by
   the NER CEGB.

 (b) 300 MW unit oil fired analogue simulator for the
   Northern Ireland Electricity Service. Commissioned
   in 1978. Engineered by the NER CEGB.

 (c) 60 MW oil fired boiler/30 MW turbine and desalin-
   ation plant digital simulator for the Bahrain State
   Electricity Directorate. Commissioned in 1979.
   Engineered by NER in collaboration with CEGB
   Computing Department.

 (d) 600 MW unit coal fired analogue simulator for the
   State Electricity Commission, Australia. Commissioned in 1980. Engineered by NER CEGB.

(Continued)

TABLE 3 (Cont.)

(e) 250 MW unit coal fired analogue simulator for the
     Huntley Power Station, New Zealand. Commissioned
     in 1981. Engineered by the NER CEGB.

(f) 350 MW coal/oil fired digital simulator, Castle
     Peak 'A' Power Station, China Light and Power Company.
     Commissioned in 1981. Engineered by the NER and
     GDCD, CEGB.

(g) 660 MW coal/oil fired digital simulator, Castle
     Peak 'B' Power Station, China Light and Power
     Company, Commissioned 1984. Engineered by the NER
     and GDCD, CEGB.

3. <u>Nuclear plant simulators for in-house training</u>
    <u>(Oldbury Nuclear Power Training Centre)</u>

(a) General purpose Magnox recirculation drum boiler
     analogue simulator. Commissioned in 1959, scrapped
     in 1974. Engineered by Elliots to a CEGB specif-
     ication.

(b) Magnox reactor analogue simulator, 5 axial mesh
     points or 5 radial/azimuthal modes. Commissioned
     in 1966. Engineered by Rediffusion to a CEGB
     specification.

(c) Hinkley Point 'B' AGR. Total plant digital sim-
     ulator. Accepted for training 1980. Engineered
     by E&T, Computing Department and GDCD, CEGB.

(d) Dungeness 'B' AGR total plant digital simulator.
     Accepted for training 1982. Engineered by E&T,
     Computing Department and GDCD, CEGB.

(e) Hartlepool/Heysham I AGR total plant digital
     simulator. Accepted for training 1982. Engineered
     by E&T, Computing Department and GDCD, CEGB.

(f) Heysham II AGR total plant digital simulator. Under
     construction. Acceptance test 1985. Engineered by
     E&T, Computing Department and GDCD, CEGB.

(g) Sizewell 'B' PWR total plant digital simulator. In
     planning phase.

Digital total plant simulators have been commissioned for Hinkley Point 'B', 1980; Dungeness 'B', 1982, Hartlepool and Heysham I, 1982. In addition a simulator for Heysham II is currently being designed and acceptance tests are planned for 1985. These are all in-house developments. The Sizewell 'B' PWR simulator is currently in the planning phase.

In addition to the above the CEGB has given considerable modelling support to the SSEB in the context of their procurement of the Hunterston 'B'/Torness simulator from Marconi Instruments.

The above historical resumé is detailed in Table 3. It should be noted that the digital nuclear simulators conserve the concept of the simulator software being upward compatible in the main frame computing architectures.

The Oldbury Nuclear Power Training Centre in Avon is connected into the main frame computer system at the Computing Bureau in London and utilises the particular computer configuration available at a given point in time, see Section 3. Given this philosophy the simulator software is conserved over the operational life of the plant for which training is required, i.e. 30 or 40 years. It should also be noted that these models were developed in PMSP and use a specialised real time integration algorithm developed in the Board. The flexibility of this approach is such that models can be sophisticated as the speed of the main frame computers allows.

Compared to models developed for purely design purposes, the training simulators required detailed models of all the mechanical and electrical auxiliary and support systems. Such systems are likely to play an ever increasing part in the more elaborate analysis of fault and operational procedures, the likely application of a plant analyser.

V. THE REQUIREMENTS AND INTERPRETATION OF DESIGN QUALITY ASSURANCE

The cost significance of poor plant availability has always been such that advanced design and manufacturing techniques have been employed in the electrical power industry. The introduction of nuclear plant required even greater attention to detail as a result of the radiation hazard and inaccessibility of many nuclear reactor components, and the requirement for detailed fault analysis. As technology has advanced more attention has been paid to underwriting design confidence levels and this has culminated in the application of Design Quality Assurance. This discipline has been mandatory in the case of the Heysham II and Sizewell 'B' designs and will be applied to the procurement of any subsequent plant.

This discipline implies that the Nuclear Installation Inspectorate can audit the Board's calculational methods and procedures and that the Board has a responsibility to audit the validity of the methods employed by its contractors. Establishing such procedures requires detailed management consideration in order to obtain the appropriate confidence levels for a limited resource and to subsequently secure this investment.

In the context of total plant modelling this discipline has been interpreted as follows:

5.1 The Quality Assurance Audit

It is necessary to provide a framework against which an audit of a given computational route can be effected. If the result of such an audit is to be satisfactory it follows that this statement is also the basis of a specification defining the requirements for the design, substantiation and reporting of the calculational route itself.

5.2 The Objective of the Calculational Route

The calculational route can be used in two basic ways. The calculations are either:

(i) exploratory - in which case no formal Quality Assurance route is required. It will be a subjective decision that the calculation is adequate for the purpose for which it is being deployed.

(ii) definitive - in this case the calculations either define, or support the definition of, operational, safety or other arguments demonstrating a viable design and operational basis for the plant over its useful life.

It follows that a definitive calculation must have a specific objective. As a generality this objective will have one of three forms as follows:

5.2.1 Operational Dynamics and Control

These calculations will proceed in best estimate data
at a subjective confidence level which is such that the operational procedure, stability margin and other consideration
is not invalidated by the uncertainty in the calculation.
The acceptability of the calculation is established by sensitivity analysis.

5.2.2 Plant Cumulative Damage

These calculations will comprise:

(a)   The dynamics of those variables, e.g. pressure
        and temperature, leading to structural damage,
        e.g. creep, fatigue, corrosion etc.

(b)   The statistical derivation of the frequency of
        the event.

(c)   The stress distribution and other considerations
        leading to the structural damage.

The calculations of (a) will proceed in best estimate
data along the lines of 5.2.1 but paying detailed attention
to the behaviour of the various structural components. In
the case of (b) operational dynamics will be predicted on the
basis of load demand and grid control requirements and a
statistical analysis of the behaviour of operational plant;
fault frequencies will be determined via event tree and fault
tree analyses. The calculations of (c) will involve detailed
theoretical analysis and/or design codes with accepted
pessimisms. The area lacking explicit definition in this
calculational route is necessarily the estimate of the frequency
of the event. This is, however, not a Quality Assurance
problem as the frequency will be monitored over plant life
and suitable corrective action taken. Nevertheless, the
design basis declaration must be realistic if plant life is
not to be curtailed.

5.2.3 Fault Analysis

This implies an ability to demonstrate that a certain
variable which may have a best estimate, peak systematic or
peak random definition, has a value not greater than some
limiting value. This may be done on the basis of a probabilistic argument, via sensitivity analysis, pessimisation of

data, or experiment. The limiting value may depend on the
frequency of the event.

*In each case it must be shown that the calculational
route is adequate for the purpose for which it is being used.*

5.3 The Calculational Route

The total calculational route may involve a number of
codes such that taken together they form an adequate definition of the system being studied. Alternatively the system
may be defined by one code. In general the QA requirements
are the same although in the former case numerical convergence
of the total calculational route must be demonstrated. For
the purposes of definition the verification requirements will
be broken down into the following areas for detailed consideration.

5.3.1 Model definition - geometric and physical definition of the plant being modelled - fault and operational
range of its defining variables - modelling assumptions and
their justification by analysis and/or experiment - derivation
of model equations from first principles - derivation of
geometric data, manufacturing tolerances - source and derivation of physical data with experimental uncertainties.

5.3.2 Numerical solution - convergence of spatial and
distributive noding - steady state solution and solution in
time; accuracy and convergence.

5.3.3 Computational veracity - simulation language and
numerical analysis packages, validity and repeatability -
reference transients, data handling and archiving programs -
computer operational software and architecture.

5.3.4 Plant and rig experimental data - instrumentation
calibration and response - filtering - recording - archiving.

5.3.5 Rig modelling.

5.3.6 Code verification - analytical solutions - code
to code comparison - rig and plant experimental verification.

5.4 Audit Requirements

The following framework is completely general and requires
to be interpreted in relation to a given calculational route.

5.4.1 Model Definition

The objective of the calculational route must be declared
(see 5.2 above). Given this declaration the plant to be
modelled can be identified as can the bounding operational
range of the main plant variables. The plant so defined must
now be expressed in mathematical form. This necessitates
making a number of simplifying assumptions as to the geometry
of the plant and the physical phenomena being invoked. In
writing the equations these shall be derived from first
principles and all simplifying assumptions shall be declared.
Wherever possible the assumptions shall be justified by
invoking arguments based on analytical and/or experimental
evidence or by sensitivity analysis. In some instances
pragmatic decisions will have to be taken due to limitations
in the available technology; if this is the case it should be
so stated. The steps being taken to remove this impediment
should be declared.

Model data will normally fall into one of three categories:
geometric, physical or generic. Geometric data should be
independently derived from the drawings by two individuals
and checked for parity. Where the form of the equations
requires a specific interpretation of the geometry this shall
be agreed and declared. For the purposes of sensitivity
analysis the tolerance on the mean data shall be stated.
Physical data will always be based on experiment. In order
to identify the data most appropriate for a given range and
application it is usually necessary to seek expert advice.
This advice should be sought and a formal consensus of opinion
obtained and recorded. Where codes are available (notably
in the reactor physics area) for the calculation and systematic condensation of data these shall be used. Those
responsible for the generation and/or support of such codes
should be required to provide an appropriate QA statement.
The recommendation to use a particular calculational route
may be by consensus of expert opinion, possibly through a
formal committee structure. Again such recommendation should
be formally recorded. Uncertainties in the data should be
declared. These are required for the purposes of sensitivity
analysis or alternatively to initiate further experimental
work where the data base is inadequate or the uncertainty
cannot be absorbed by the design. By generic data is meant
library routines which have general application and have been
called to service a particular calculation. The thermodynamic
and transport properties of fluids, or nuclear cross-section

data would be examples of such an application. The user of
such library routines should ensure that an appropriate QA
route exists for any such routine deployed in his calculation.

Whenever it is expedient to proceed without a viable QA
route for declared data items, or there is an inability to
justify modelling assumptions, a statement shall be given of
the steps which are being taken to resolve the difficulty.
The likely significance of the uncertainty shall be stated in
order that suitable strategic action can be taken e.g. by the
design of confirmatory commissioning experiments.

5.4.2 Numerical Solution

When formulating the equations of 5.4.1 it will be necessary to spatially distribute equations e.g. in the axial and
radial sense, or alternatively to distribute nodes in pipework
and other physical systems. All such noded calculations shall
be shown to have converged numerical solutions by the experimental choice of the number and distribution of nodes and/or
the data weighting deployed. In the case of total plant
modelling in the simulation language PMSP, the accuracy of the
steady state solution should be established by checking that
the time derivatives are zero and by running a null transient.
In the case of fixed time step integration algorithms it shall
be demonstrated, by numerical experiment, that the solution
has converged for the particular forcing functions being
deployed. In the case of predictor corrector algorithms
with variable step length, it may be adequate to argue
convergence by using tight convergence criteria such that
the calculation is adequately converged for any forcing
function. Such arguments should be stated and justified.

If calculations are effected in other than double
precision it shall be demonstrated that the use of single
precision does not invalidate the numerical methods and that
the calculation yields a converged solution for the forcing
functions being deployed.

5.4.3 Computational Veracity

All computer codes should be provided with a set of
reference calculations so designed that they embrace all
likely applications. In the case of dynamic calculations
the reference transients should be chosen in such a way
that the full range of time constants is excited. A precise
definition should be given of the computer operating system,

compiler and machine architecture or other consideration
necessary to completely define the computational route used
for the generation of the reference transients. If a given
code is subsequently executed in a different computational
environment and/or on a different machine it should be shown
that the reference transients are satisfied. If the code is
modified or extended the reference should be regenerated and
the process repeated.

In the case of the simulation language PMSP the origin-
ators should check the validity of the translator, sort
algorithm and other numerical procedures by executing a code
or codes which have a stand alone capability, e.g. KINAGRAX,*
establishing a one to one correspondence between such a code
and a PMSP equivalent, inclusive of the integration algorithm,
and ensure parity between reference transients. Where it is
possible to establish analytical checks on integration,
linearisation, frequency response and other algorithms this
should be done. Whenever a new release of the program is made
the various test cases should be re-run and the verification
route thus re-established and declared. A similar verification
procedure should be deployed and any analytical package e.g.
Fourier analysis. All verified codes must be conserved in a
secure and incorruptible store and diverse storage must be
provided.

Adequate procedures must exist for change control,
archiving of code releases and recovery via some librarian system.

5.4.4 Plant and Rig Experimental Data

Any experiment which is used for model and/or code
validation purposes or in justification of modelling assumptions shall itself have an adequate Quality Assurance route.

Both plant and rig steady state and dynamic behaviour
have to be interpreted through a complement of instrumentation.
The steady state calibration and dynamic response of all such
transducers is to be determined with estimates of calibration
error. This information is to be determined over the valid
operational range of the transducer and this range shall be
declared. In many instances the mounting detail of the
instrument will itself be of importance. Hence the above
requirement must relate to the variable being identified and
not just the response of the transducer itself.

----

* See Table 8.6.1.

The output from any transducer may be amplified, possibly
filtered and in any case will be recorded and/or logged.
This process in itself can introduce errors. Hence the significance of this route should be determined in terms of its
frequency response, filtering algorithms, time skewness,
logging frequency and any other relevant consideration.

All the above information and rig experimental output
must be stored in computer readable form in precisely defined
physical units and archived for the life of plant or code to
which it relates.

5.4.5 Rig Modelling

As a general proposition rigs will approximate real
plant in terms of geometry and/or their operational range.
They may be based on scaled models or analogues of the process
involved and/or attempt to conserve dimensional similarity.
In every instance the experimental modelling assumptions should
be declared and justified in relation to the experimental
objective. In general it will be necessary to model the rig
mathematically in which case all the requirements of Sections
5.4.1, 5.4.2 and 5.4.3 must be observed. In order to minimise
the experimental uncertainties the geometric and physical
data of the rig should be specifically determined and subsequently deployed in the mathematical model of the rig.
Detailed attention must be given to rig heat losses and other
phenomena which relate to the rig rather than the plant being
approximated.

5.4.6 Code Verification

Wherever possible design codes should have their validity
established by as many of the following techniques as possible:

(a)   Analytical solutions.

(b)   Code to code comparison.

(c)   Code to rig verifications.

(d)   Code to plant verification.

Each of these techniques will be considered individually;
if all four are deployed the confidence level will be
maximised.

(a)   In specialised cases it will be possible to set up an <u>analytical solution</u> defining the linear response and/or stability of a given plant detail. Given such a solution it is possible to examine the convergence of spatial noding for a linear solution and compare the linear time response and frequency response with that derived independently from a non linear model. Other linear techniques e.g. Fourier analysis could equally well be deployed.

(b)   In general codes will be either 'reference' codes i.e. those codes based on the most rigorous physical definition of the phenomena being studied, or 'engineering' codes, i.e. those codes giving adequate answers in a defined frame of reference, but making various simplifying assumptions in the interests of minimising solution costs. Wherever possible all such engineering codes should be compared in the steady state, and if appropriate, for bounding transients. Necessarily the interpretation of any such <u>code to code</u> comparison must be subjective but the solution obtained from the engineering code must be demonstrably adequate for any given application.

(c)   <u>Code to rig verification</u> should proceed using rig and rig model by linear analysis in the frequency domain and by comparison of non-linear responses in the time domain. Given that adequate agreement is obtained, modelling aspects of the rig should be stripped away and the relevant and verified physics conserved to the plant model. If sensitivity analysis suggests that adequate agreement can be obtained only by invoking additional phenomena then the restrictive range of application of the code should be declared, as should the work programme designed to remove the restriction.

(d)   <u>Code to plant verification</u> should in principle only proceed when the requirements of 5.4.6 a, b & c have been met as appropriate. In general the linear and non-linear analysis of 5.4.6 c can be deployed. Total plant model validation should only proceed after a high confidence level has been established in the code modules

representing the major plant items. It is necessary to pay particular attention to the dynamic
behaviour and non-linearities of all plant actuators,
control and protection systems, initiating open
and closed loop testing, as necessary, to establish
an adequate confidence level. No statistical
methodology exists at the present time for defining the error in amplitude and time of a given
variable and hence the acceptance of a given
dynamic prediction relative to experiment is
necessarily subjective. Clearly both the originator of the comparison and the auditor must be
experienced in this discipline and guided in their
judgement by a framework of steady state and
dynamic sensitivity analysis. This should embrace
geometric, physical and experimental data and
modelling assumptions. For minimum error in
establishing the maximum value of a given variable
for a particular fault, or other consideration, an
experiment approximating the fault scenario must
be designed. Under these circumstances the uncertainty reduces to establishing the experimental
error in the transducer identifying the particular
variable under consideration.

All calculations associated with sensitivity analysis,
code to plant comparison and the design of experiments must
be recorded, archived and reported, and a definitive release
of the model and its data set secured for posterity. All
subsequent calculations should proceed using the latest
release of the code which must be the subject of the requirements of Section 5.4.3.

5.5 Cost Significance

Clearly the cost significance of promoting such a discipline is considerable. This being the case it is necessary
to conserve all stand-alone codes and the elements (macros)
of total plant models and the models themselves in a secure
and disciplined environment.

The way in which the macro philosophy is being developed
is a consequence of the above QA considerations and this is
explained in detail in Section VII.

VI. THE SIMULATION LANGUAGE PMSP AND
RELATED DESIGN CODES

From Section III it will be realised that the Board has
taken an interest in the design and application of simulation
languages since 1964. Early Dungeness 'B' digital modelling
was effected in DSL90 (1966) and subsequently transferred into
CSMP marketed by IBM. Extensive application of the latter
code resulted in a decision for the Board to develop and
provide its own simulation language. This was duly done and
released for general application within the Board in 1970.
It was realised at an early date that plant models would grow
rapidly to a size which would not allow of hybrid computer
solution. The EAL 8800 system introduced in 1969 was phased
out in 1975 and during this period considerable emphasis was
put on the development of fast integration algorithms in
anticipation of a wholly digital computer environment. This
development was very successful and continues in an attempt to
keep pace with modelling sophistication.

Currently the Plant Modelling System Program (PMSP)
provides the main computing facility in the CEGB for the
analysis of power plant models, covering a wide range of
applications for both nuclear and conventional plant. The
code has been developed to provide a robust and mature system
offering a very reliable service to some 300 users throughout
the Board as well as other major U.K. companies e.g. NNC,
NEI Nuclear Systems, Babcock Power and the AEA.

Plant models are written by engineers in the special
PMSP high level simulation language which are presented to
a "Translator" program for processing into FORTRAN statements.
Subsequent steps (compilation, linking and execution) allow
the model to be assembled (including selection of an
appropriate numerical integration algorithm) into an
executable program to run a simulation. Special facilities
have been built to link with other systems and to provide
extensive pre- and post-processing facilities.

To aid portability and allow a degree of machine
independence, the PMSP program is written in a special high
level language called MORTRAN. There are already versions
of PMSP which have been run on ICL, VAX and SEL computers, as
well as the usual IBM compatible machine. A CDC version is
also available.

Experience with PMSP has been such that it has become
essential to the Board and the U.K. Power Industry. To
secure the system and ensure a satisfactory advisory service,
a team of analysts has been assigned within the Computing
Bureau who are able to give a comprehensive service to users
whilst carrying out an identified development programme.

As a result a system incorporating a wide range of
enhancements was released in 1984. Effort is also being
directed towards the provision of MACRO facilities and
associated archiving features to enhance quality assurance
procedures and to allow the rapid assembly of a given plant
geometry from an inventory of plant modules.

In-house and collaborative work on the interfacing of
other modelling packages e.g. ALADDIN for Control System
Design, NUMEL for Boiler Calculations, is being undertaken
to extend the options and area of application of the PMSP
code.

Internal algorithms are constantly under review and updated to ensure maximum efficiency and minimum solution time.
It is recognised that the responsiveness of interactive
services to PMSP and the expansion of the range of plant
models will necessitate an order of magnitude improvement
in current machine processing performance.

PMSP is the base language for both design simulation
studies (non-real time) and for conventional and nuclear
training simulators (real time), enabling steady state and
dynamic analysis to be carried out with the same system
without changing the model. The development of total plant
models for design and safety investigations is the major
force behind the need to secure the future of PMSP. Whilst
the user interaction with design type models does not warrant
the sophistication of Control Room instrumentation replication,
it will necessitate the development of more advanced
processing facilities to enable users to selectively operate
and examine the conditions and behaviour of plant.

6.1   Introduction to PMSP

6.1.1   The Modelling Process

The modern engineer makes wide use of mathematical
modelling to supplement the physical or engineering means
of finding a solution to a practical problem. It allows

the feasibility of alternative designs to be tested before construction and can remove the need for expensive prototype equipment. Physical experimentation can be particularly costly when destructive testing is involved, and in some cases - where operational problems arise with a power station, for instance - experiments on the plant itself might present a risk to power transmission, or even to the safety of the plant itself. In these circumstances, it is often necessary to perform the same experiments on an accurate simulation model. In the case of nuclear plant fault analysis there is no other way of demonstrating the viability of the plant design and its protection system and thus establish the safety case leading to a plant licence.

A common feature of the mathematics used in these situations is that most physical processes may be represented by a set of algebraic and first-order ordinary differential equations. This mathematical representation is important because once the form of the equation is recognised, standard methods of solution can be used, and the experience gained on one job can be applied to another. A digital computer solution is required by the scale of the modelling involved. The solution of algebraic equations usually involves the iterative method whilst that of differential equations requires the method of numerical integration. The Plant Modelling System Program (PMSP) has been designed to allow the user to solve systems of equations of both types conveniently, and in the shortest possible time.

6.1.2 The Plant Modelling System Program - PMSP

This is a high-level language which enables dynamic systems to be simulated and analysed. The language is FORTRAN-based and uses double precision arithmetic where necessary - which in practice means almost everywhere. Any process capable of being described as a set of algebraic and differential equations can be modelled using PMSP.

The simulation model is developed by preparing PMSP structure, data, and control statements to correspond with the user's concept of the underlying physical principles involved in the study of a particular problem.

Specifically designed to meet the needs of scientists and engineers who require digital simulation mainly as a design tool, PMSP is flexible and easy to use. Preparing

input for the system takes time, but no great programming
knowledge is required, and the user is free to concentrate on
the phenomena being studied rather than the mechanism being
used to implement the study. This means he is able to give
more attention to whether or not the model accurately
represents his concept of the phenomena in the light of his
theoretical and practical experience.

The technology is simple and readily understandable by
the engineer, so that the input - basically in the form of
first-order differential and algebraic equations - usually
looks very much like the original mathematical definition of
the problem. Each facility is invoked by a simple command
and the numerical methods used were developed particularly
for their reliability and efficiency.

Although the construction of a block diagram - or a set
of algebraic and differential equations - representing the
phenomena to be studied is an important first step in the
successful use of PMSP, it is not essential to complete this
stage entirely before making the first simulation run. In
fact it is quite instructive, and may well save time, to
build up a large model by developing it section by section.

Another important feature of PMSP is that data entry
and information required for output are simplified by means
of free format data and output control statements. With few
exceptions, data and control statements may be entered in any
desired order and may be intermixed with structure statements.
The PMSP language processor (translator) can be used to
automatically determine the sequence necessary to establish
the correct information flow. The translator transforms the
model description into a FORTRAN subroutine which is compiled
and then executed with the selected numerical algorithm.

The language offers a great many simply-invoked functions
which commonly occur in continuous processes, including time
delay, dead space, limiter functions, function generators,
and first and second order lags. However, its main appeal
lies in the range of mathematical and analytical techniques
available to the user.

The basic PMSP set includes integrators, comparators,
function generators, and switches. These are supplemented
by FORTRAN library functions such as cosine, square root,

and absolute value. If the user requires other functions for
a particular problem, these may be defined either as FORTRAN
subprograms or, more simply and directly, through a macro
feature which allows a larger functional block to be constructed out of elementary statements or existing functions.

Output options include printing of variables in standard
tabular or equation format, point plotting in line graphical
form at selected increments of the independent variable,
usually time; and direct digital incremental plotting. In
addition output can be written to disc and analysed through
the graphics package VISION.

The numerical facilities offered include:

(a)   Automatic steady states

        The user is not required to integrate from an
        arbitrary initial state in order to determine the
        steady state of the model; instead two reliable
        methods of directly calculating the steady state
        are provided.

(b)   Fast Integration

        A number of integration routines are provided
        including some specifically designed to deal with
        models with widely differing time constants.

(c)   Automatic Linearisation

        Realistic models are non-linear and often incorp-
        orate empirical data. Linearisation gives access
        to many useful analysis and design techniques.

(d)   Frequency Response Analysis

        The frequency response of any model variable with
        respect to any other variable or parameter can be
        calculated. The program enables loops to be
        broken so that, for example, open loop responses
        may be calculated from a closed loop model. Time
        delays are represented explicitly, and the results
        may be displayed in the form of Bode, Nyquist
        and inverse Nyquist plots.

(e)   Eigenvector Analysis

The eigenvalues (poles) of a model and their
corresponding eigenvectors can be calculated, thus
enabling the origins of oscillatory modes, in-
stabilities, and stiffness to be determined.

(f)   Control System Design

PMSP interfaces with a system of programs contain-
ing a variety of design facilities for single and
multivariable control systems - ALADDIN.

(g)   Input/Output

Input and output are simplified by means of free-
format data entry and user-orientated input and
output control statements.

(h)   Transparency

Only one formulation of the model is required and
all algorithms operate in a 'transparent' manner
on the same model. Each facility is invoked by
a simple command.

(i)   Model Checking and Precision

Special precautions have been taken to ensure that
incorrect or inaccurate results do not arise as a
result of the pathological numerical properties
of the model. Similarly the model is specified
in double precision, and, with the exception of
integration, results are computed to machine
precision. Attention has also been paid to the
inclusion of diagnostic aids and automatic
checking features to aid model building.

(j)   Program Limitation

The maximum model size is limited only by the
available computing resources and each program
employs the minimum resources required for its
solution.

6.1.3 PMSP Application

The development of a simulation study using PMSP may be divided into four main stages although at times these stages may overlap. The first stage involves the specification of the modelling equations and the basic data, with stage two defined as the conversion of these equations and data into a program suitable for input to the PMSP system. The third stage is concerned with debugging and testing the program before the fourth stage, which involves using the model to obtain engineering information. The blurring of the boundaries between these stages occurs because it is sometimes necessary to alter the basic equations to overcome problems encountered in the numerical integration of the model equations, or because the model may consist of a number of sections and while certain sections are being developed, others are being test run in isolation.

Before starting stage one it is essential, if the right interpretation is to be put on the results of a simulation, to study the physics of the process to be modelled and identify those important points in the characteristic behaviour of the plant that must be included in the model. It is also necessary to recognise those points of the process under study which are not so important, but which enable simplifying assumptions based on sound engineering reasoning to be made.

When defining the problem in mathematical terms, the idea is to express the phenomena under investigation either as a set of first order ordinary differential and algebraic equations, or as functional blocks with interconnections, or a mixture of both.

Continuous system simulation languages – specifically designed for use by those who are not computing specialists – are particularly appropriate in the plant modelling applications area. PMSP includes a basic set of functional blocks which may be used to represent the component parts of the mathematical model. Under certain conditions the program permits the use of FORTRAN-type statements. This is very useful for performing logic and branching, as well as special input/output if required. More complex sophisticated modelling techniques can be carried out by the extensive use of FORTRAN.

The steady state finder plays a significant part in the

testing and debugging of the model as well as being used to find steady states for the whole model. Very few models are developed using PMSP which do not require steady states to be found.

6.1.4 The Characteristics of PMSP

The four main requirements of any problem-solving system are first - and most importantly - that it should be robust; that it should require minimum intervention in respect of the provision of such items as error criteria, scaling of problems, number of iterations allowed, and so on; that it should include comprehensive error diagnostics expressed in terms of properties of the simulation model, and finally, the system should be as efficient and economical as the previous requirements allow.

Over the years, experience has shown that the algorithms used are very robust for the models presented to PMSP though not necessarily in any other context. PMSP requires no provision of error criteria, with the single exception of the numerical solution of differential equations, where it seems unavoidable that users should state the accuracy required. For all other applications, answers are computed to the limiting accuracy for the machine and algorithms used. For practical applications, the extra work involved in proceeding from a "reasonable" accuracy to the limiting accuracy is not so great and provides greater confidence in the solution obtained.

Where difficulties are experienced in problem solving with PMSP, errors in one of three areas may be the cause. It must be remembered that the mathematical model does not accurately describe the physics of the problem for a particular representation, and to achieve a valid interpretation of the results, the user must always be as aware as possible of the approximations made in a particular model. It is important to record carefully any conscious approximations made during development to ensure that their possible effects may be considered during any later evaluation.

Errors involving data may be of two types. Either accurate data cannot be obtained, or the data cannot be matched to corresponding coefficients of the mathematical equations since, as stated above, there cannot be an exact correspondence between the physics and the mathematics. This may lead to some equations having to be redefined.

Mistakes may also occur at the processing stage ; to ensure accuracy of solution, great precision is necessary coupled with the appropriate numerical techniques. PMSP provides the means of obtaining accurate results, and makes full use of the latest numerical methods; it is updated and modified as and where necessary.

6.2 PMSP Numerical Algorithms

Models will consist of large sets of non-linear ordinary differential and algebraic equations. Typically there may be as many as 500 or more ordinary differential equations and several thousand associated algebraic equations.

The equations are non-linear, because nature is non-linear, and realistic models always take account of this. The o.d.e.s are usually "stiff" with a stiffness ratio of about 10^9 and are usually quite strongly cross-coupled, making the automatic separation of "stiff" and"non-stiff" components very difficult. There is an important minority of problems which are not "stiff" in the usual sense but have solutions of approximately fixed frequency - usually 50 cycles per second. As often occurs in many types of simulation model, the Jacobians associated with the differential equations are sparse - typically 5% dense - but unfortunately exhibit no symmetry numerically or positionally. The equations are almost always discontinuous in the first derivative because of discontinuous forcing functions, and in higher derivatives because of the use of function fits to empirical or experimentally obtained data.

6.2.1 User Requirements from PMSP

In terms only of numerical algorithms, users require the following facilities, all of which are available in PMSP:

(a) Numerical solution of ordinary differential equations, both stiff and non-stiff.

(b) Steady state solutions
By this is meant the quiescent state of the model, i.e. that state at which time derivatives are zero or acceptably small. Almost all studies, at some stage, require the steady state to be found, the model solved for a short while at this state i.e. a null transient, and a disturbance then applied. Accurate steady state solutions are essential so that cause and effect can be distinguished.

(c) Access to linear analysis techniques
     Although models are almost always non-linear, valuable
     insight can be gained through the application of linear
     techniques.

     In this context, PMSP provides:

     (i)   Numerical linearisation of the model about an
             operating point (usually the steady state condition).

     (ii)   Eigenvalue analysis of the linearised system
             A knowledge of the eigenvalues can answer certain
             questions concerning the dynamic stability of the
             model.

     (iii)   Frequency response calculations
             This facility is very widely used by engineers in
             control system design. Output from frequency
             response calculations is usually requested in the
             form of Nyquist, Inverse Nyquist or Bode plots.

     (iv)   Steady state sensitivity analysis
             This provides an estimate of the condition of the
             steady state solution with respect to the model
             data. This device is especially useful when the
             error tolerance of the data is known.

     (v)   Origins of instability, stiffness or oscillatory
             modes
             It sometimes happens that the user's model exhibits,
             for example, dynamic instability as indicated by
             at least one eigenvalue of the linearised system
             having a positive real part. This knowledge is of
             limited use unless that part of the model which
             is, in some sense, the "cause" of the instability
             can be detected. A partial answer to this problem
             is to examine the associated eigenvector, due
             allowance having been made for the effect of
             different physical units. This eigenvector then
             "points", sometimes with remarkable accuracy, to
             the required area of the model.

             PMSP can perform all the above tasks on simple
             commands from the user and using exactly the same
             definition of the model. The fact that the model
             need not be changed according to algorithmic
             requirements is of great importance to the user.

In terms of performance of numerical algorithms,
it is found that users usually have the following
descending order of priority:

Robustness

This is easily the most important requirement.
Answers must be obtained if possible.

Minimum Intervention

In respect of provision of such items as error
criteria, scaling of problems, number of iterations
allowed etc.

Comprehensive Error Diagnostics

All such diagnostics must be expressed in terms
of properties of the simulation model. As a
simple example, an error message which says
"INTEGRATOR VARIABLE *FLUX* IS MISSING" is far
more valuable to the user than the equivalent
"COLUMN 12 OF THE JACOBIAN IS NULL - FACTORISATION
IMPOSSIBLE".

Efficiency

This comes last in order of priority, although
it is of course essential to see that simulation
is as cheap as possible, consistent with (a),
(b), (c) above.

Where needed, PMSP scales a problem automatically. For
example, in solving non-linear simultaneous equations (steady
state problems), the equations are scaled to remove the effect
of different physical units. When eigenvalues are needed,
the matrix in question is balanced beforehand and whenever
sparse LU factorisations are to be performed, the matrix is
equilibriated.

It is difficult to assess the efficiency of the algorithms
provided. For a new algorithm to be considered as a replacement for an existing one, it would have to be shown, over
a period of time, to be overwhelmingly superior. Examples of
replacements which have taken place are the use of sparse
instead of full matrix techniques and the inclusion of several
"stiff" integration algorithms.

6.2.2 <u>Description of the Numerical Algorithms used</u>

Whenever, in the following text, reference is made to sparse matrix algorithms, it will imply the use of the algorithms described in Curtis and Reid (1971) [6].

(i) <u>Numerical solution of ordinary differential equations</u>

As described earlier, almost all systems presented fall into the category known as "stiff", and here three different variable step algorithms are available.

The first is an implementation of Gear's algorithm[1]. The strategy as proposed in that reference is all but useless for applications to large sparse systems such as those for which PMSP is used. The following is a brief summary of the changes made to that scheme:

(a)   All matrix operations are replaced by sparse matrix code. The need for this change is self-evident for systems of hundreds of differential equations, known in advance to be sparse.

(b)   Gear's policy of recalculating the Jacobian numerically whenever the order of method being used changes or there are problems with convergence, is disastrous for large systems. In general, we adopt the policy that if the current iteration matrix solves the non-linear algebraic problem at each time step in a reasonable number of iterations, then we retain it. When there is a problem with convergence, we have found that this is almost always because the current steplength and/or order has changed greatly from that in use when the iteration matrix was last calculated. It is rarely because the Jacobian elements themselves have changed too much. We therefore keep a copy of the last Jacobian calculated in core, in sparse form, and whenever the rate of convergence is too slow or divergence occurs, the iteration matrix is simply reconstructed by combining the stored Jacobian, the <u>current</u> steplength and <u>current</u> order. In practice, little more is required than a re-scaling of the diagonal elements of the iteration matrix followed by a "fast" sparse LU decomposition. The decomposition is a "fast" one because it assumes that the Jacobian structure has remained unaltered and that the pivotal sequence found to be best at the start of the integration is still adequate.

The iteration matrix, referred to above, is of the form:

$$D = I - ahJ \qquad (6.1)$$

where h is the steplength, a depends on the order and J is the Jacobian of the system of differential equations. I is the identity matrix.

Thus when divergence occurs, a and h are updated but not J. D is stored in core as:

$$D = (I/ah - J) \qquad (6.2)$$

so that only the diagonal terms of D need to be changed. D is then decomposed into its triangular factors, maintaining sparsity, and these triangles over-write the old ones. This procedure usually produces sufficiently improved convergence but other recovery procedures are then tried if this is not so. If this fails to produce convergence, then usually it implies that the structure of the modules has changed and J needs to be recomputed, a very expensive process. Whenever LU factorisation takes place, a check is kept on the pivotal growth of errors and if this is too great, a new pivotal sequence is calculated but this necessity is found only rarely. It is commonly found that during a transient solution, only one Jacobian need be calculated over many hundreds of seconds, re-scaling being adequate to produce convergence. It has been found too, that successive corrector iterations respond well to an Aitken (delta)^2 acceleration process.

The algorithm just described is invoked by the simple command METHOD MERCURY.

The second method available for stiff systems is one due to Fowler and Warten [2]. This is a non-linear explicit scheme which fits a simple exponential function locally to the solution. It is designed for systems having widely spaced eigenvalues of the local Jacobian. Even when the eigenvalues are clustered, the scheme works well but in the presence of oscillatory modes, efficiency seems to decrease rapidly.

The most generally used algorithm available for stiff systems is an unusual implementation of the implicit trapezoidal rule. As is well known, the simplest method of solving the non-linear algebra at each time step is the method of successive substitution so that the iterative sequence to generate the solution at the (n+1)th time step is, with obvious notation

$$\underset{n+1}{\overset{(i+1)}{y}} = \underset{n}{y} + \frac{h}{2} \{ \overset{\cdot}{\underset{n}{y}} + \overset{\cdot \, (i)}{\underset{n+1}{y}} \} \qquad (6.3)$$

This method, of course, diverges rapidly for stiff problems for usable values of h. If, however, three successive members of the sequence (6.3) are saved and an Aitken (delta)² process applied, componentwise, to these three vectors, the "accelerated" vector so produced is, under certain circumstances, a good approximation to the solution of the algebraic problem. If this vector is then used as the first of a new triad obtained from (6.3) and the acceleration applied again, convergence of the accelerated vectors is often rapid. The method works best when the "stiffest" eigenvalue lies on the real axis, well separated from the others. This kind of distribution is, fortunately, frequently found in models presented to PMSP. The method is a variable step method possessing the great advantage of requiring no storage of Jacobians and thus makes light demands on core and has very low overheads. It is robust in the presence of discontinuities, presumably because of its low order (2) and its being a one-step method. This algorithm is invoked by the command METHOD WARP2.

For non-stiff problems, users can choose between a 4th order variable step Runge-Kutta scheme or Gear's non-stiff option (Gear 1971) [1]. Little effort has gone into the incorporation of good non-stiff methods because very few systems encountered are non-stiff.

(ii) Steady State Problems

These require the calculation of the quiescent state, i.e. that stage at which time derivatives are acceptably small. This implies the solution of a square system of non-linear algebraic equations. For less than about 50 equations, an implementation of a Broyden rank-1 update scheme [3] is used. In conjunction with an automatic scaling procedure to remove the effect of physical units, this scheme has proved remarkably robust over many years of use. For very large systems, however, it is essential to exploit sparseness and for this an implementation of Schubert's method [4], coupled with sparse matrix techniques, is proving equally

effective. Up to 600 non-linear equations have been solved
with this method. It too includes automatic scaling and
requires the storage in core of a Jacobian approximation and
its triangular decomposition, both in sparse form. For both
these algorithms, solutions are found to the limiting
accuracy using the technique described in Section (viii).

(iii) Numerical Linearisation of Non-Linear Equations

For access to linear systems techniques, the Jacobian of
the system must be computed accurately. In computing accurate
numerical partial derivatives, the usual problems of conflict
between heavy cancellation of leading digits and truncation
error is solved by using central difference approximations
applied iteratively to achieve the limiting accuracy. The
process is accelerated by Richardson extrapolation.

(iv) Computation of Eigenvalues of the Linearised System

To obtain information concerning the stability and/or
stiffness of the dynamic system, some of the eigenvalues of
the Jacobian (calculated as in Section (ii)) are needed.
Very few eigenvalues are required from the matrix, these being
at either end of the whole spectrum. Those with most positive
real part give information concerning stability, those with
most negative real part give information concerning stiffness.
For this problem, an effective method is the following
procedure:

(a)   Scan the matrix and remove physically ignorable
        elements.

(b)   Balance the matrix.

(c)   Reduce the matrix to upper-Hessenberg form using
        stabilised elementary similarity transformations.
        Call this matrix H.

(d)   Scan the sub-diagonal elements of H for zeros and
        if any, split the matrix into groups of smaller
        matrices.

(e)   Using Hyman's method, find the determinant and
        its first and second derivatives with respect to
        L, of:

$$D = (H - LI) \tag{6.4}$$

where L is the current approximation to an
eigenvalue. Use these in Laguerre's method for
computing the roots of polyomials.

(f)   Use root suppression - see Wilkinson, [5] to
        remove the possibility of converging more than once
        to an eigenvalue already found.

We have found that this method has the great advantage
of allowing eigenvalues to be extracted at will from either
end of the spectrum. This can be achieved simply by making
the initial estimate for the root to lie to the left ("stiffness") or right ("instability") in the complex plane of all
eigenvalues. The algorithm described is extremely robust
and has been used successfully on very many problems over a
number of years. Its main disadvantage is its comparative
slowness. Interestingly, although convergence with Laguerre's
method is guaranteed only when all roots are real, we have
never detected any difficulty in obtaining convergence for
complex roots, but only in the rate of convergence when these
roots lie close to others.

(v) Frequency Response Calculations

The frequency response of a stable system is the response
the system attains after an infinite time in response to a
sinusoidal forcing function of fixed frequency. To calculate
the frequency response we have to find the LU decomposition
of matrices of the form.

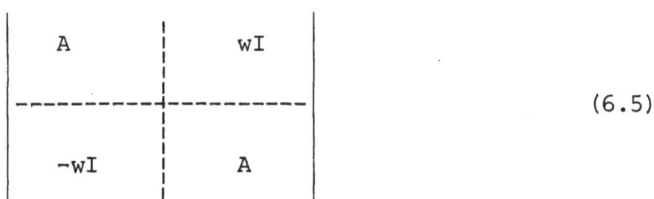

$$(6.5)$$

where A is square and sparse, I the identity matrix and w the
frequency in question. Often A will be nearly singular and
w very small and in that case, iterative refinement of the
solution is essential. Since it is not known in advance how
badly-conditioned A is with respect to inversion, iterative
refinement is always applied for all A and w. Sparseness is
exploited by performing a sparse factorisation of (6.5).The
matrix has to be factorised for possibly hundreds of values
of w for the same A, because users usually require the

frequency responses to construct Nyquist, Inverse Nyquist or
Bode plots. We have found it best to solve for values of w
in decreasing orders of magnitude. The pivotal sequence
determined for the first value of w is used for all subsequent
values of w. If iterative refinement fails, then a new
pivotal sequence is calculated and the calculation proceeds.
Hence, almost all decompositions are "fast" in the sense of
Curtis and Reid [6]. It is rare for more than one pivotal
sequence to be needed.

(vi) <u>Sensitivity Analysis</u>

It is important to know the effect on numerical solutions
of perturbations in the model data, especially when the error
tolerance of the data is known. We have found that it is
sufficient to do this for steady state calculations, since
sensitivity here usually indicates that most other calculations
which may be requested (e.g. numerical integration) will
probably also be sensitive. If numerical calculations are
shown to be highly sensitive over the known tolerance, then
the validity of the whole simulation study must be called into
question. The steady state sensitivity analysis may be
found as a frequency response (see previous section) for w
very small.

(vii)<u>Origins of Instability, Stiffness and Oscillatory Modes</u>

As mentioned in Section 6.2.1, the knowledge that a
model is, for example, dynamically unstable is of little use
unless the cause of the instability can be detected. Almost
always, dynamic instability results from a user coding error
but we have found that the eigenvector corresponding to the
"unstable" eigenvalue indicates, often with remarkable
accuracy, that area of the model in which the coding error
lies. The effect of physical units is removed from this
eigenvector and its non-zero components are given to the
user as the physical names of the offending differential
equations. The eigenvector is found by performing inverse
iteration on the matrix:

$$(H - LI) \qquad\qquad\qquad (6.6)$$

where H is the accurately computed Jacobian of the system
(Section (iii)) and L the "unstable", "stiff", or
"oscillatory" eigenvalue, as required. If L is complex, real
arithmetic is still used by performing inverse iteration on

a matrix twice as large as that shown in (6.6). Sparse LU
factorisations are performed on (6.6) and using the sparse
triangles so formed, a series of forward and back substit-
utions is performed, starting with an arbitrary vector, in
the usual way. The effect of physical units can be removed
by performing a simple diagonal similarity transformation
on H where the diagonal terms are simply the steady state
values found as in Section (ii).

(viii) <u>Convergence Criteria used to Terminate Iterative</u>
       <u>Sequences</u>

PMSP users are usually unwilling to provide termination
and accuracy criteria for iterative numerical calculations.
A general technique is used by PMSP to terminate all iterative
sequences. This method provides:

(a)   the limiting accuracy in numerical procedures
       for the given algorithm and machine precision.
(b)   independence of machine precision.
(c)   greater confidence in numerical answers.
(d)   no user intervention.

The technique has been described by Wilkinson (1965) [8],
and is based upon the idea that if an iterative sequence of
approximation z_i, i the iteration number, is converging then:

$$\left| z_{i+2} - z_{i+1} \right| < \left| z_{i+1} - z_i \right| \qquad (6.7)$$

until the limiting precision for the algorithm and machine
has been reached. Initially (6.7) may not be true until the
sequence z_i "settles down" and so the covergence criteria
used are:

$$\left| z_{i+1} - z_i \right| < e \text{ and } \left| z_{i+2} - z_{i+1} \right| < \left| z_{i+1} - z_i \right|$$

$$(6.8)$$

e is a "generous" error tolerance, dependent on the algorithm
used. Iteration continues until (6.8) is satisfied.

There is a possible danger in using (6.8) that excessive
computer time may be used if convergence is very slow. However,
in view of the criteria described in Section 1, the risk is
in general worth taking. The criteria (6.8) are used in PMSP

to terminate iterative sequences for steady state solutions, numerical linearisation, eigenvalue calculations, frequency response calculations (iterative refinement), and "origin" calculations (inverse iteration).

6.3 PMSP - Developments in Integration Algorithms

It is recognised that the efficiency of numerical integration algorithms is of the first importance in the field of large scale simulation. Current developments of MERCURY (see earlier), based on extensive experience of the application of the algorithm, have yielded another algorithm, known as 'M2' which has a much simpler strategy than the parent algorithm. It is more robust in the presence of discontinuities, and achieves this by recognising that discontinuities provoke an inefficient over-response from numerical algorithms, and inhibits this by restricting the order of the method used and by ignoring 'obvious' excessively large internal error estimates.

An alternative approach being investigated is based on Porsching's algorithm [7] and will be known as 'M3'. It has long been recognised that the extensive 'overheads' in M2 and Mercury were largely a result of the need to iterate within the algorithm at each time step. These overheads become heavier still when model discontinuities are encountered, this being an extremely common occurrence in total plant modelling. M3 removes these overheads by not iterating at all, and yet still retains the stability properties of the classical implicit methods, a feature thought to be impossible until recently. Besides having the immense advantage of being non-iterative and yet stable, an equally important benefit is that the numerical expressions evaluated at each time step yield an extremely effective method for detecting discontinuities. The ability to detect discontinuities easily and the non-iterative nature of M3 both promise to yield an extremely powerful algorithm which may well change the hitherto accepted ideas of the economics of large total plant modelling.

VII. THE CONSERVATION OF VALIDATED PLANT MODELS - ARCHIVED PLANT MACROS

7.1 Introduction

From the point of view of minimising the manpower resource associated with code development validation and the

maximisation of design confidence levels, it was found
desirable to develop a particular plant modelling strategy.
Hitherto the deployment of such an approach has been seriously
hindered by the available computing power but whilst this is
still a consideration, it no longer dominates considerations
relating either to practicability or cost.

In the past modelling has proceeded on an ad hoc basis.
Models were constructed to answer specific questions for
minimum computing cost in the available computing environment,
and at a subjective confidence level. As plant models have
become larger, unconstrained by the computing environment, the
development cost of such models has been largely associated
with manpower and this is clearly an incentive to standardisation. Furthermore the application of Design Quality
Assurance is a positive disincentive to modify plant models
once these have an established verification and validation
route.

For the above reasons a particular plant modelling
strategy has been developed. This strategy, whilst it is
not able to handle all classes of calculation, is not particular to a given class of plant and allows the conservation
of all common plant elements e.g. once-though boiler models
have a common equation set for AGR or CFR plant.

In developing this strategy it has to be accepted that
there are still limitations in modelling capability. These
arise from three considerations, namely:

(a) The potential degrees of asymmetry in the model e.g. the
     number of boilers; PWR models can readily handle 4
     boilers, whilst the AGR, in the case of Heysham II say,
     has 24 main boilers and 24 shutdown decay heat boilers
     and not all of these can be modelled. Some plants have
     multiple turbine systems; these can be modelled but have
     to be considered in relation to the number of boilers;
     4 boilers and 2 turbines in the case of the PWR is
     demonstrably feasible.

(b) Dimensionality. In principle plant is 3-dimensional but
     in the context of total plant modelling most plant
     components are modelled as one-dimensional axially dis-
     tributed systems. Currently it is not feasible to model
     2- or 3-dimensional reactor or boiler models in the
     context of total plant calculations.

(c) Extreme complexity of analysis of certain low probability
     events, e.g. loss of coolant accidents involving core
     reflood. Such calculations may require specialised codes.

Implicit in the above statements is the concept of software being upward compatible in developing computer architectures. It is no doubt feasible to develop software specific
to a given specialised architecture, but this is seen as being
counterproductive to the long term development of the modelling philosophy.

Below is set out the modelling strategy being developed
in the Plant Kinetics Group of the Boiler Branch within the
Generation Development and Construction Division, and the
relationship between this strategy and the development work
in the Board's Laboratories and Computing Bureau.

It also considers the application of Board developed
design methods by its agents and contractors and the contribution which they and others can make to the proposed strategy.

The proposition developed herein does not have universal
application, in that it would not, for instance, embrace
3-dimensional reactor or boiler dynamics as discussed above,
although there is a hierarchial relationship between the
latter codes and those which are the subject of this statement. The proposition as presented is necessarily restricted,
but nevertheless embraces a vast modelling area as declared
below.

7.2 Functional Responsibilities of the Plant Kinetics Group
     in GDCD

The Plant Kinetics Group has design responsibility for
the operational dynamics and control of conventional and
nuclear plant and is also required to provide a numerical
service to the Division. In principle this suggests an
infinite class of problems but in practice most of the problems
can be resolved in the context of total plant modelling e.g.
safety analysis or the dynamic input to plant cumulative
damage calculations. The Group also has sub-let responsibilities for the mathematical modelling associated with plant
training simulators, e.g. Heysham II/Torness and Castle Peak
'B'. Such modelling should conserve the design models.
Hitherto this has not been practical as the complexity of the
design models militated against real time solution. However,

the development of computer architectures is such that in the
near future this will no longer be the case. If the design
models are conserved to the training simulator there is
necessarily a considerable saving in model development and a
significantly higher confidence level if the design quality
assurance route is conserved.

All total plant modelling proceeds in the simulation
language PMSP; see Section VI above.

7.3 Modelling Disciplines

It is a Project requirement for Heysham II/Torness and
Sizewell 'B' that Design Quality Assurance shall operate.
This is no more than a formalisation of good engineering
design practice and as a consequence is equally relevant to
conventional and nuclear plant. This discipline requires
that the total plant model and/or its constituent parts should
be validated against reference models, as well as rig and
plant experimental data, and the validation conserved. It is
here suggested that the discipline for achieving this should
be as follows:

7.3.1 Definitions

For the purpose of this discussion the following
definitions have been adopted.

(a)   By validation is meant the critical comparison
of the results of a more sophisticated code, with
experimental rig results, with plant experimental
data or with all three.

(b)   By verification is meant the reporting route
whereby the equations and assumptions of the
model, the data derivation, the model validation
route and the user manual for the code are
explicitly defined.

(c)   By conserved is meant the retention of those
modules or macros which have a common mathematical
representation for a given plant component which
is relevant to more than one type of plant e.g.
once-through boiler modelling for Wylfa (Magnox)
or the CFR boiler. These macros require characterisation by the appropriate data set.

See Section V for a more detailed discussion of Design
Quality Assurance.

7.3.2 Computational Methods

The British Power Industry as a whole is now carrying out
total plant modelling in the simulation language PMSP. This
allows of unrestricted model size and has in its library fast
integration algorithms. Whereas in the past various groups
within the Industry developed specialised techniques and codes
to answer specific questions, it is now practical for most
steady state and dynamic calculations to proceed in the one
Code. This has the advantage of conservation of data and model
structure and the potential advantage of conserving the validation and verification routes.

Previously reference codes were written to define in detail
the physics of a particular component of the total plant and
simpler engineering codes were written and validated against
the reference model. Computational speeds are now such that
this artifice is becoming unnecessary and it is possible to
code the reference model directly into PMSP. For instance
this has been done in the context of the Magnox reactor code
KINAX, and the AGR reactor code KINAGRAX.   Clearly such a
procedure minimises the uncertainty and the manpower resource
associated with the model validation route. In other areas
the Board has developed specialised design suites at very
high confidence level which tend to make obsolete numerous
other design codes. An example of this is the once-through
boiler code NUMEL. This provides a dynamic boiler code
which has a steady state solution which is adequate for most
design purposes and links with the code DYMEL to provide
information regarding boiler hydrodynamic stability, thereby
providing a self-consistent statement of boiler behaviour.
NUMEL has also been successfully implemented in PMSP. All
three codes are examples of a PMSP macro.

7.3.3 Modelling Macro-Structure in PMSP

For the purposes of explanation an example is developed
based on the existing experience in respect of Heysham II.
However, the principles can be equally well deployed in the
context of design models for nuclear and conventional plant,
or of nuclear or conventional plant modelling for training
simulators. In the latter case it may be necessary to
conserve a simple model definition in the interests of real

time solution and cost, both restrictions becoming of little
significance in relation to developing computer technology.

Heysham II/Torness total plant modelling is based on
the Hinkley Point 'B' work with a modified control structure.

The Hinkley Point 'B' model currently (1984) has a
modular structure which resulted from the need to write the
model as a series of FORTRAN subroutines, which could be
written by more than one individual, and solved using the
numerical methods available in PMSP. Since this time the
size restrictions of PMSP arrays have been removed and it is
intended that the Heysham II model should be written wholly
within PMSP. It is proposed that the model should be broken
down into a series of 'macros' defining *all the plant components
from relief valves to reactors*. The precise definition of a
macro will evolve as a matter of experience and in any case
will be conditioned by the validation route available.

The difficulties in definition may be illustrated by
consideration of a start up/standby feed pump macro. This
comprises an induction motor, hydraulic coupling,suction pump
and main feed pump. Each one of these components could be a
macro in its own right or alternatively they could collectively form a standby feed pump macro. Even if the composite
macro were adopted it is essential to agree and conserve the
modelling adopted for the individual elements. This is to
prevent arbitrary differences in modelling which could arise,
for instance, in the case of Dungeness 'B' circulators which
are also driven by an induction motor through a hydraulic
coupling. However, when validating a macro its boundary
conditions will depend upon whether the validation is against
another code, real plant or a rig. In real plant cases the
boundary conditions are rigidly defined by the plane of the
installed instrumentation. The position of the installed
instrumentation will be such that it is the standby feed
pump macro which is validated whilst the components of the
macro will be valid by collective inference. It is also
necessary to recognise that the validation of the composite
macro will require the modelling of any pipework up to the
plane of the experimental instrumentation and the detailed
modelling of the instrumentation itself. The experimental
macro when validated, will have stripped away from it the
pipework and instrumentation modelling to yield the verified
macro. Clearly if insufficient attention is given to
modelling the instrumentation the resultant errors will be

included in the confidence level of the verified macro and
this confidence level will be diminished in an unidentifiable
manner.

It may not be possible to model all aspects of the plant
on the basis of its physics and geometry. In such a case it
may be necessary to deduce the dynamics from experimental
data. This proposition is better understood through the
exploration of a given example.

The boiler gas outlet mass flow (in the case of an AGR)
leaves the main boiler at a particular temperature, passes
through the dead thermal capacity of the shutdown decay heat
loop, through the insulated void to the circulator and thence
to the void under the diagrid. (For the purposes of discussion the various cooling flow splits are here ignored). The
attenuation in temperature and the mean path length of the gas
cannot be directly calculated due to the complicated distribution of masses and lack of gas dynamic definition. It is
therefore necessary to deduce this model from experimental
data. In order to define the generality of such a model it
has to be fitted over the power range to as many initial
conditions and forcing functions as possible. Again it is
important to ensure that the boundary conditions of this
deduced macro are accurately defined by the macros either
side of its boundaries. If inadequate attention is given to
this point the errors of adjacent macros will be swept up
into the implied physics of the deduced macro.

From the foregoing arguments specific definitions
materialise:

7.3.3.1 Experimental Macro

This is defined as the aggregate of various plant
components contained between boundaries defined
by experimental instrumentation. It will include
the dynamics of the instrumentation and any pipework (or other entity) between the boundary of
the component and the plane of the instrumentation.

7.3.3.2 Composite Macro

This is defined as a combination of a number of
plant components which collectively define a plant
element the boundaries of which are capable of
experimental definition.

7.3.3.3 Component Macro

This is defined as the smallest unit of plant
component which has a separate physical entity.

7.3.3.4 Deduced Macro

This is defined as a plant entity which cannot be
directly characterised by its physics and geometry
in that they are indeterminate and have to be
deduced experimentally.

An illustration of this structure is given on
Figure 1.

Further definitions could well be required in the context
of simulator modelling.

7.3.4 Modelling Micro-Structure

A number of component macros are one-dimensional axially
distributed models e.g. the reactor, boiler, steam-mains and
scatter plug/ballast weight. In current practice the partial
differential equations (pde) are defined and then discretised
in space to yield a set of ordinary differential equations.
These equations are subsequently solved using the numerical
integration methods available in PMSP. Using PMSP facilities
and the Method of Lines it is possible to work directly with
the partial differential equations. Such a pde definition
forms a micro structure, the definition of which should be
conserved from three-dimensional reference models. Again,
this is best understood by example. Consider, by way of
example only, the three-dimensional AGR reference model SKARK.
This has a macro -structure which is defined by the pdes for
the one energy group neutron kinetic and thermodynamic model
at a point. In order to simplify the verification route
this pde definition should be conserved in the one-dimensional
reference code KINAGRAX and in any PMSP equivalent. In the
interests of minimising the solution time, it may be necessary
to simplify the pde statement in PMSP. If this was the case
as many of the equations as possible should be conserved and
any restructuring should limit the uncertainty to a specific
area and hence isolate out the validation uncertainty.

In the present example this might require the condensation of the fuel model from three rings to a lumped fuel

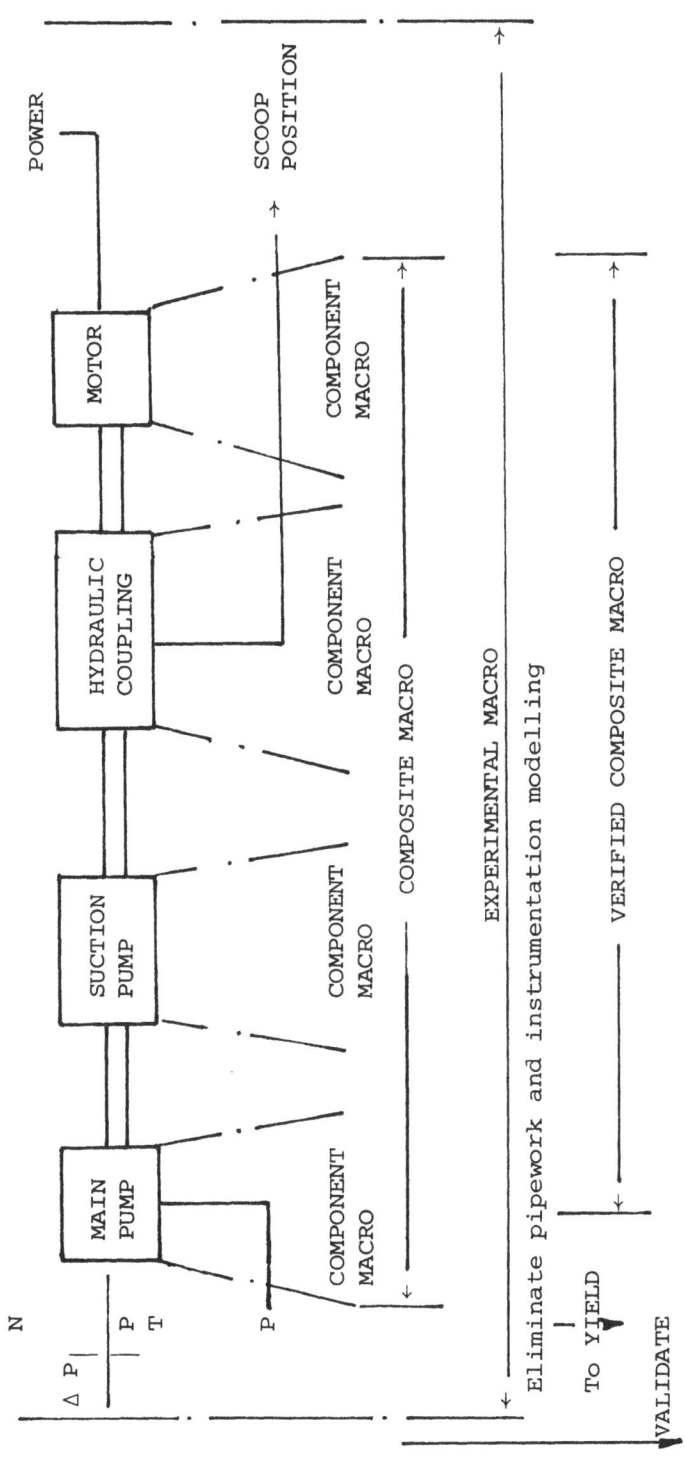

Figure 1.   FAMILY OF "MACROS"

model whilst conserving the sleeve and bulk moderator
definition. The validation uncertainty is now only that
associated with the Doppler feed back in the k-infinity equation.

It should be further noted that as a matter of principle
it is not possible to separate out uncertainties in data from
those of the equations. Hence, if a set of data is derived
by fitting to experimental measurements the data should be
deployed only within the equation definition used for its
derivation. An example of this would be the fitting of the
fuel temperature coefficient in KINAGRAX against Hinkley Point
'B' experimental data. By extension it is the KINAGRAX
equations which should be conserved in PMSP.

7.3.5 Computing Connotations

The fundamental purpose of the macro and micro structure
is to further the development of a detailed verification route
and its subsequent conservation. It does, however, allow of
further computational developments which lead, in effect, to
a higher level simulation language. Such a language is
currently being developed and this will allow the automatic
encoding of macros thus enabling the engineer to concentrate
more on the solution of design and operational problems rather
than the development of the computational route.

7.3.6 Reference and other Codes not in PMSP

In order to establish the validity of a plant design,
steady state and dynamic codes will be deployed which are not
written in the PMSP simulation language. These codes may have
no direct bearing on total plant dynamics but are a necessary
part of the verification route. An example of this would be
the two-dimensional boiler codes based on CROSSMIX, the one
dimensional version of which should be validated against NUMEL
and thus with the 12-plane moving boundary model in PMSP.
Other reference codes such as the boiler design suite NUMEL
and the three and one-dimensional reactor codes SKARK and
KINAGRAX are more directly involved in the hierarchy of the
verification procedure in that they will provide evidence as
to the validity of the relevant component macros in the PMSP
total plant modelling structure. See Figures 2 and 3.

All such codes require a verification route which should
be referenced from the code/macro library.

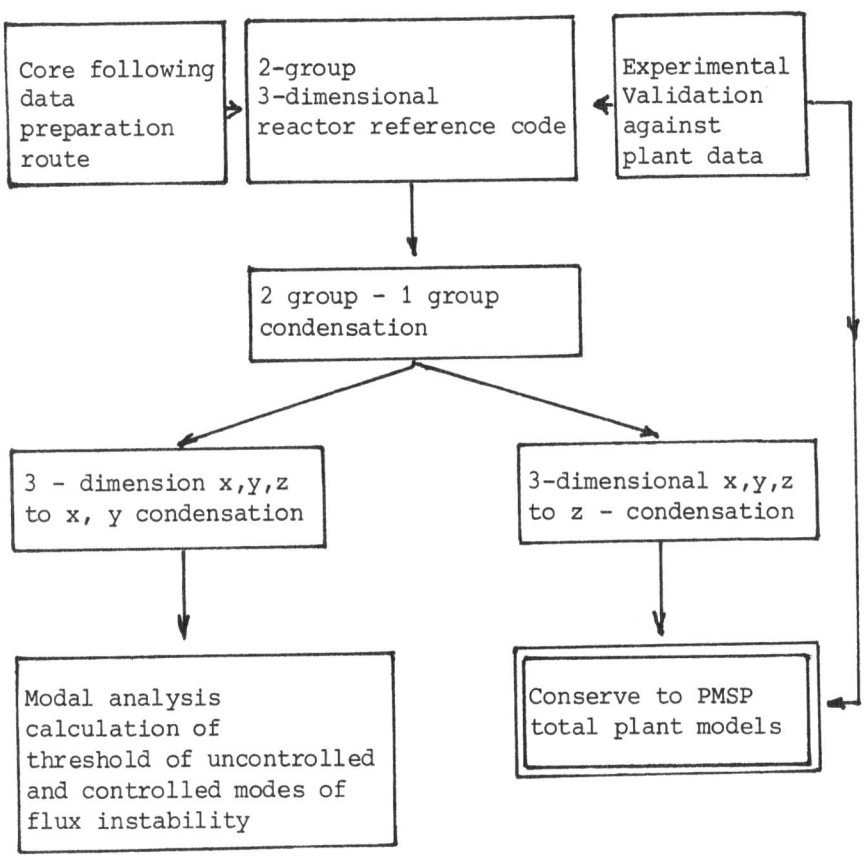

Relevant codes:

Magnox   AKTS, SKIP, KINAX   Data route:
                                     ADOS
AGR   AKTS, SKARK, KINAGRAX

PWR   AKTS

Figure 2.

DESIGN QUALITY ASSURANCE
CONSERVATION OF THE HIERARCHIAL VALIDATION ROUTE
AGR & PWR EXAMPLE. GENERIC DEFINITION
REACTOR PHYSICS MODELLING

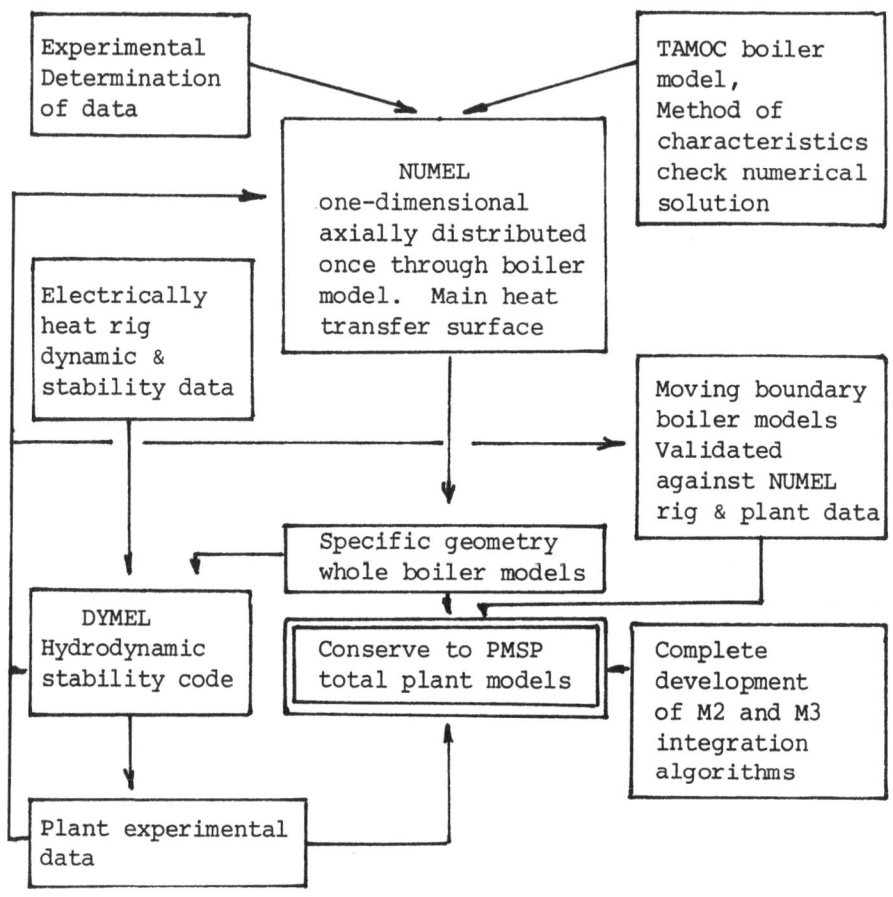

Figure 3.

DESIGN QUALITY ASSURANCE
CONSERVATION OF THE HIERARCHIAL VALIDATION ROUTE
AGR & CRF EXAMPLE. GENERIC DEFINITION
NUMEL — REFERENCE BOILER MODELLING

7.3.7 The Plant Modelling Library and its Verification

A library is being constructed which will contain within
it the precise definition of the component, composite, deduced

and _experimental_ macros each of which would have associated
with it a verification route the purpose of which is to
provide a statement of the validity of the model and its data.
It will be appreciated that the macro definitions of section
7.3.3 are collectively all embracing in that the _composite_
macro can develop to the level of a total plant model. Furthermore the verification route will include a structured
hierarchial argument which will embrace the verification route
of any reference model having separate application or invoked
to establish the validity of a corresponding macro.

In the remainder of this section the philosophy and
thinking that underlies the library is described.

7.3.7.1 Code Verification Route

It is required that every code/macro has supporting documentation as follows:

(a) A complete statement of the code/macro
equations, their derivation from first
principles, and the assumptions on which
they are based.

(b) The derivation of plant data, its accuracy
and special characterisation e.g. the weighting of three-dimensional reactor data sets
to a one-dimensional axial array.

(c) A user manual for any code/macro having stand
alone application and similarly for any
composite macro approximation to a total
plant model.

(d) A model/macro validation route comprising:
(i) Model systematic reduction from a 2/3-
dimensional calculation having its own
validation/verification route.

(ii) Code to code validation.

(iii)Rig to model validation.

(iv) Real plant to model validation.

Clearly it is not possible in the short term to construct a library which comprises the full verification route for all codes or macros. This being the case it is proposed that the library should be developed as follows:

Conserved codes/macros

Clearly a calculational route already exists for the majority of design calculations. These calculational routes are expected to be technically sound, will be used in good faith, but in general will not have a verification route of the form identified above. The latter will have to be provided. It is therefore proposed that the library should originally consist of the conserved codes/macros i.e. it will be a statement of current practice.

Verified codes/macros

As verification routes are provided for the various conserved codes and macros in the library the definition of the code/macro will be updated from conserved to verified. Subsequently all calculations where appropriate, will proceed using the verified definition and will be fully supported by the proposed documentation. In so far as full verification may be impractical at a given point in time for want of experimental data, or some other reason, this will be explicitly identified as will the proposed corrective action. It is thought that such a procedure will have a valuable infeed into the distribution of plant instrumentation and the design of commissioning experiments.

Default codes/macros

In order to study certain aspects of plant behaviour in the absence of a final design it will be necessary to model anticipated plant components either in PMSP or say, in a subroutine of NUMEL. These might take the form of transducers or actuators, for example. As such the equations and data will be defined in the library as default macros i.e. they will be used in default of an explicit design statement. They will, of course, ultimately, be replaced by a component macro. This is an example in which the PMSP component macro has replaced a default macro and has an application outside PMSP.

Simulator macros

Owing to the current computing constraints on real-time simulation it has been found necessary to develop simplified models which by definition cannot conserve the verification route but nevertheless should be conserved to the macro library. Such macros should be updated to the verified definition as and when computing power allows.

7.3.8 Change Control

When a calculation is carried out it will be necessary to archive the source code, data set and forcing functions. The source code and data set will clearly have a quite explicit definition and an explicit verification route, although the conserved definition may be used in the short term. Design quality assurance will require that this particular verification route be conserved in the library, even though the conserved code may change its status to a verified code or the latter may be updated i.e. all issues of the verification route must be retained in the library.

7.3.9 Macro Modelling Disciplines

In so far as various engineers will be contributing to the macro library it has been necessary to set up and agree with interested parties various modelling disciplines e.g. macro coding structure, notation, normalisation, sub-routines for thermodynamic and transport properties, general use of non-dimensional functions for heat transfer and pressure drop calculations, generalised look-up tables for multi-dimensional surfaces for pump and circulator characteristics and so on.

These disciplines will be formalised and identified in the plant modelling library.

7.3.10 Code/Macro Archiving and Security

Clearly having established the code/macro library it will be necessary to archive this within a secure computing library environment. Equally it will be necessary to secure all experimental data whether from rigs or plant and all source code, data sets and forcing functions wich may have been deployed within the verification route. Such information will have to be archived indefinitely and certainly over the

life of a given plant from its design inception to decommissioning, i.e. some fifty years.

7.4   The Significance of the PMSP Macro Library to the Design of a Plant Analyser

The application of a plant analyser within the design and
operational environment carries with it the probability of
model modification, enhancement and extension. Given the
above philosophy it is unlikely that modification to the major
plant macros would be required and in all probability any
modification would relate to details in the support systems
e.g. the chemical and volume control system in the case of a
PWR, or the control or protection systems. Clearly any such
model update can most readily proceed in a simulation
language e.g. PMSP, since no particular attention has to be
given to the numerical solution of the updated equation set.
Furthermore if the modification can invoke a macro which is
already available in the library it can proceed at the plant
component rather than the equation level.

This modelling philosophy is currently in an advanced
state of realisation. Reference models are available in PMSP
in macro form as are other plant modules. Three sophisticated
total plant models, for Sizewell 'B' (PWR), Hartlepool and
Heysham I (AGR), are currently being converted from subroutine
to macro form and a fourth for Heysham II is in an advanced
state of development. The advanced integration algorithms
M2 and M3 and PMSP structural and library archiving development are all reaching fruition. It is confidently expected
that the availability of plant experimental data, currently
accruing from the commissioning of Hartlepool and Heysham I
will lead to the verification and validation of the constituent
macros as well as the total plant models themselves. The
validation of the appropriate macros will then be conserved
to Sizewell 'B' and Heysham II (AGR) e.g. the construction
of the turbine feed train macros will be such that they can
be conserved to any plant whether nuclear or conventional.

Given that such a macro library is established it
should clearly be conserved throughout the Board and the
British Power Industry. By extension it could be the subject
of international collaboration and in this context it is
interesting to note that Combustion Engineering Power Systems
propose to offer a PMSP bureau service in the USA.

VIII. THE CONVERGED MODELLING PHILOSOPHY
AND ITS RELATIONSHIP TO TRAINING
SIMULATORS AND PLANT ANALYSERS

8.1 Background

It will be realised from the detailed information given
above that the CEGB has been comprehensively involved in the
design and safety analysis of nuclear plant since 1957 (Section
III), and with the design and procurement of plant training
simulators since 1959 (Section IV). This work has proceeded
against a background of extensive plant procurement and in an
environment in which the Board Laboratories have explored the
behaviour of plant components through large rig facilities,
the determination of physical data and the development of
fundamental theories relating to the various physical phenomena
appertaining to plant design. In addition it has been explained
how an adequate design feedback has been assured through plant
commissioning procedures and the high speed logging of plant
experimental data from an elaborate instrumentation provision
(Section II). The Board has worked, and is working at an
advanced level of technology in all the relevant disciplines,
and in so doing has extensively deployed analogue, hybrid
and digital computers for the purposes of numerical analysis.
As a result of the application of Design Quality Assurance
(Section V) and the employment of the simulation language
PMSP (Section VI), advanced mathematical and computing
techniques have and are being developed which allow of plant
calculations proceeding at high confidence level.

8.2 Computing Considerations

Because of the enormous financial investment in software
which apertains to the Board's design and licensing activities
it is the Board's policy that this software should be upward
compatible with the computing architectures being used for
in-house training and design purposes. For a given plant this
software is expected to remain viable over the period from
design to decommissioning, some 40-50 years. This precludes
the use of specialised architectures and their software. The
currently deployed interactive editing facility TSOEASY and
the output graphics system VISION would thus be conserved.
In effect it would be transparent to the user on which machine
his code was executed. Machine implementation would then be
purely a matter of turn round and cost, but solution at very
low values of the ratio of computer time to real-time would be
impossible.

8.3 Software Considerations

Clearly the Board has a large range of codes to cover the needs of design, licensing, commissioning and operational activities. A number of these codes fall into the categories identified in 7.1 as incapable of inclusion within total plant models at the present time e.g. the 3-dimensional reactor codes SKIP (Magnox), SKARK (AGR), AKTS (PWR); 2-dimensional boiler codes, based on CROSSMIX, PWR detailed LOCA codes such as TRAC. However, these codes have specialist applications to symmetric reactor rod faults, boiler orificing and loss of coolant accidents and may well be of value in a plant analyser environment, but within the Board the bulk of total plant numerical analysis would take place using the simulation language PMSP. As explained in Section VI, this simplifies the coding and makes available to the engineer a highly sophist-icated numerical analysis package which thus enables the user to concentrate on the resolution of design problems. Given that the macro library discussed in Section VII is taken to its logical conclusion this will enable modelling to proceed at the plant component, rather than the equation level, and allow of the flexible extension of models to explore design innovation. Such a modelling facility obviously allows of the modelling of any plant type. In practice Magnox, AGR, PWR and CFR nuclear plant design models, simulator models and conventional plant models are all available within PMSP. If exploration suggested that a particular plant problem could be resolved by extension of the control structure this could readily be done using the interactive design suite ALADDIN. This contains all the necessary single and multi variable linear control system design theorems in the S and Z planes and would allow of a new topology being rapidly developed, the non-linear aspects of this design subsequently being explored heuristically. File handling and other facilities have been developed which link this design suite to PMSP via the open loop transfer function or state space matrix.

8.4 Input Output Facilities

Existing input and output modelling facilities require further development. Currently inter-active editing is carried out using the editing language TSOEASY via keyboard VDU facilities with hard copy printing capability. This system is fully developed and wholly viable. A development area is the organisation of the code interface to enable the

changes in code state and the application of forcing functions
to be the more readily implemented. This problem has been
addressed for training simulators, (in the context of the
Tutors Desk), but not generalised to the design environment.

8.4.1 Code Input Interface

PMSP input is currently organised as follows. A subroutine DATVAR is used for data manipulation. Default data
sets exist and would normally be deployed. There will be
more than one such data set in order to handle plant ageing
e.g. fuel cycle considerations. All control variables available
to the operator are brought together in subroutine DESK. This
makes available logic states, control system set points, auto/
manual stations, valve positions and so on. Currently control
system modules are invoked with the associated plant. However,
the development of the macro structure will isolate out the
control systems and work is in hand to emulate the control
system implementation in CUTLASS, the on-line computer language
which will be used for control system implementation.
Rationalisation will take place as the macro structure is
implemented. Plant accident states are normally pre-identified
in relation to the model noding e.g. in the case of core
depressurisation accident the break occurs at a particular node
and picks up the appropriate choked orifice equations. No
attempt at generalisation has so far been considered. The
introduction of plant forcing functions can also be difficult
and may require a detailed knowledge of the model to overwrite
control functions. This may lead to superfluous integrators
and special branching. If the introduction of forcing
functions is to be generalised this area requires detailed
consideration.

8.4.2 Code Output Analysis

Currently editing is carried out in TSOEASY and the
code is offered up to batch execution and the output, pre-
identified, written to disc. This allows of automatic scaling
and the output is analysed using the VISION graphics package.
Two- and three-dimensional plotting facilities are available,
in colour, if required. Facilities are available for
plotting frequency response data, Nyquist diagrams and so on.
Steady state and dynamic mimics have been developed for plant
operational purposes but have not been deployed for design.
Potentially an enormous volume of data is available for analysis.
The significance of this data can be drawn to the engineers

attention by flagging that constraint boundaries have been excited, margins to trip eroded and so on, but no attempt has so far been made to devise 'expert', or other methods of data analysis and reduction.

8.5 Considerations Relating to Speed of Execution

In order to obtain a feel for the execution time and values of beta (the computer to real-time ratio) for various codes the reader is referred to Tables 8.1, 2 and 3. The most robust integration algorithm available in PMSP is currently WARP2. Table 8.5/1 gives values of beta for Heysham II and Hinkley Point 'B' total plant models for various plant transients run on the IBM 3081. Values of beta range from 0.36 to 6.49. The conclusion is obvious, namely that a reduction in beta is required. This can be obtained in two ways either by further improvement in fast integration algorithms or by use of more advanced machine architectures or both.

For complex numerical reasons it has been found necessary to develop a special integration algorithm if the reference boiler code NUMEL is to be conserved to PMSP. This has been done, the integration algorithm being known as M2 and is the subject of further development, see Section 6.3.

It will be seen from Table 8.5/2 that values of beta vary between 0.22 and 0.46 for various forcing functions when applied to the boiler model alone.

Prior to the development of M2 an improvement over WARP2 was obtained by an interim development known as MERCURY. Table 8.5/3 compares values of beta for WARP2, MERCURY and M2 for various plant models and transients. It will be seen that for other than the moving boundary boiler model factors of 2 to 5 are possible between WARP2 and M2. In the case of the reactor model with direct digital control the maximum integration time step cannot exceed the sampling time. Had this not been the case an order of magnitude improvement on WARP2 would have been possible.

This work is being consolidated to yield an integration algorithm to be known as M3. It is this algorithm which will be developed to be wholly robust. Marginal improvements are expected relative to M2; any further improvements come from machine speed.

TABLE 8.5/1

REPRESENTATIVE TRANSIENTS RUN UNDER THE WARP2 INTEGRATION ALGORITHM
TOTAL PLANT MODELS FOR HEYSHAM II (HEBDYM) & HINKLEY POINT 'B' (HINDYM)
VALUES OF BETA FOR VARIOUS TRANSIENTS RUNNING ON THE IBM 3081

Code	Transient	Simulated Time /s	IBM 3081 CPU Time /s	Beta
HEBDYM13 (199 integrators)	Load ramp 80% to 45%	600	213.28	0.36
HEBDYM13 (199 integrators)	Reactor Trip	228	420	1.84
HEBDYM21 (254 integrators)	Reactor Trip and decay heat boiler start-up	182	1181	6.49
HINDYM5 (410 integrators)	Quandrant Trip	1800	4071	2.26
HINDYM5 (410 integrators)	Reactor Trip	192.2	531	2.77

Beta = CPU Time
       Transient real time

TABLE 8.5/2

DEVELOPMENT OF THE INTEGRATION ALGORITHM M2
REFERENCE BOILER CODE NUMEL RUNNING IN PMSP
VALUES OF BETA FOR VARIOUS TRANSIENTS RUNNING ON THE IBM 3081

Transient	Number of Axial Mesh Points	Simulated Time /s	Beta for 10^{-3} relative error	Beta for 10^{-4} relative error
Step reduction in feed flow	30	100	0.25	0.28
Ramp in inlet enthalpy	60	100	0.36	0.46
Hartlepool boiler trip sequence	30	800	0.22	0.25

Beta = $\dfrac{\text{CPU Time}}{\text{Transient Real Time}}$

NUMEL is a one-dimensional axially distributed once-through boiler model.

TABLE 8.5/3

COMPARATIVE VALUES OF BETA FOR VARIOUS CODE RUNNING ON THE IBM 3081 FOR VARIOUS TRANSIENTS UNDER THE INTEGRATION ALGORITHMS WARP2 AND MERCURY AND THE DEVELOPMENT ALGORITHM M2

Code	Integration Algorithm	Transient	Simulated Time /s	CPU Time /s	Beta
Hartlepool total plant model HARDYM304	WARP2 M2	Step change in load demand	500 500	2088 908	4.18 1.81
9-plane reactor, scatter plug and T_2 control	WARP2 MERCURY M2	Gas flow reduction from 100% to 80% over 10 s	250 250 250	153.9 39.5 29.0	0.616 0.158 0.116
12-plane moving boundary boiler model BOILMAC5	WARP2 MERCURY M2	Dynamic boundary conditions corresponding to a reactor trip	250 59.2 250	61.84 299.5 67.79	0.247 5.06 0.271

Beta = $\dfrac{\text{CPU Time}}{\text{Transient Real Time}}$

In the latter context the Computing Bureau will be
carrying out bench mark tests on the CRAY1, CDC-CYBER 205
and FPS 164. No scaling factors are currently available.
Whilst significant reductions in execution time are expected
it is highly improbable that very low values of beta can be
approached for models of more than say, 500 integrators.

8.6.1   The Nuclear Plant Design Analyser

This definition has the implication that betas less
than 1.0 are not necessary and hence advice to operating
plant in real time is not possible for fault recovery strategies. For this specification the plant analyser is envisaged
to have the following form.

(a)   Hardware Environment

A generalised computing environment would be established
comprising input stations in the form of a keyboard VDU
and touch screen facilities communicating with a front-end
processor. In the case of the prototype this would be so
organised that the front-end processor could communicate
with a main frame system. The object of this is to allow
any code to run in the design analyser environment. The
analyser when deployed in a given organisation may use a
dedicated machine as required by its established practices.
Output would be via banks of colour VDUs with hard copy
facilities.

(b)   Software Environment

(i)   An interactive editing environment is already
available. Mimic facilities would be provided to
ease code utilisation. These would be required
in the context of emulating desk and panel controls,
automatic control stations, forcing function input
and nodalisation for fault implementation. It is
believed that this interface is capable of complete
generality to all types of plant modelled.

(ii) Plant Models
Some 90% or so of plant analysis can be effected
in total plant models. The Board would recommend
the conservation of the Design Quality Assurance
route, the use of the simulation language PMSP and
the advanced macro structure module library, and

the associated control system design package
ALADDIN. The hardware environment as defined in
(a) allows the deployment of any other design code
e.g. for 3-dimensional core and reactor fault
analysis. Clearly any code for any type of plant
can be run in this environment. The value of beta
is unimportant but should be as low as practical
for reasons of cost. If a dedicated machine was
justified cost would not be a consideration. The
equipment should be operated against a criterion
of maximum useful utilisation.

A list of relevant codes is given in Tables 8.6/1,
2 and 3 for AGR, PWR and CFR codes. In the case
of the PWR many of the codes relate to the NSSS
only, have no modelling of the balance of plant
and inadequate modelling of other areas e.g. the
containment system and the chemical and volume
control system. In general, and certainly in the
case of the PWR models, they would require extension
to include additional systems such as those identified above. This will necessarily increase code
solution times. Because of the commercial confid-
entiality of plant data sets and the proprietary
nature of certain codes, these will have to be
stored in a secure manner under password control in
the case of a prototype analyser made available for
use by, or demonstration to, external organisations.

(iii) Output Facility and On-line Data Analysis
It is envisaged that output to VDUs would be via
graphic packages, plant steady state mimics and
dynamic mimics. Whilst graphics and plant state
mimics can be generalised at some point, on-line
analysis is likely to become plant type specific.
There is no point in running codes at very low
values of beta i.e. for values less than 0.1,
unless the analyst is given a powerful on-line
analysis facility which provides detailed and
immediate comprehension of the plant fault condition,
the probable, if not the specific cause, and likely
corrective strategies. Such a facility is probably
the key to the successful application of the plant
analyser and must be developed. It is likely
that it can only be developed in a plant type-
specific manner.

TABLE 8.6/1

AGR DESIGN CODES

DESIGN CODES HAVING DIRECT APPLICATION TO THE 'PLANT ANALYSER FOR DESIGN APPLICATIONS ONLY'

1. Total Plant Models

Code Name	Code Definition	Code Application
(a) Hinkley Point 'B' HINDYM 1,2,3,4 PMSP	Turbine feed train, alternator, grid, reactor, boiler(s), feed systems, dump systems, control and protection systems	Operational steady state and dynamic analysis, control system design, fault analysis
(b) Dungeness 'B' DUNDYM 1,2,3,4 PMSP		
	1. Point reactor, one boiler 2. Point reactor, two boilers 3. Axially distributed reactor, one boiler 4. Axially distributed reactor, two boilers	
(c) Hartlepool HARDYM 1,2,3,4 PMSP		
(d) Heysham I HEYDYM 1,2,3,4 PMSP		
(e) Hunterston 'B' HUNDYM 1,2,3,4 PMSP		
(f) Heysham II HEBDYM PMSP	As (a), (b), but being developed in the macro structure form	(c) and (d) currently being converted to macro form (1986).

2. Once-Through Boiler Models

(a) NUMEL	Reference contraflow heat exchanger model. Has been configured to numerous plant specific once-through and drum boiler configurations	Detailed steady state and dynamic boiler analysis, static and hydrodynamic stability analysis
Also available in PMSP		
(b) DYMEL	Links to NUMEL for hydro-dynamic stability analysis	Hydrodynamic stability analysis
(c) BOILMAC7 PMSP	12-plane moving boundary boiler model. Adaptable to various boiler configurations	Operational and dynamic analysis

3. Reactor Codes

(a) AKTS	3-dimensional 2-group reactor steady state and kinetics program. Provides a data preparation route, 2 group to 1 group condensation and reduction to 2-dimensional x, y and 1-dimensional z models	3,2 & 1-dimensional steady state and dynamic analysis e.g. asymmetric rod faults, on load refuelling, control system design, gas mass flow faults etc.
(b) SKAR K	3-dimensional 2 group reactor steady state and kinetics program	As 3(a)

(Continued)

TABLE 8.6/1 (Cont.)

Code Name	Code Definition	Code Application
(c) KINAGRAX Also available in PMSP	1-dimensional (z) 1 group dynamic program	Symmetric reactivity and flow faults, control system design, axial flux stability etc.
(d) 9-plane reactor model PMSP	As 3(c) with simplified heat transfer modelling	As 3 (c)
(e) ADOS	Nuclear data library steady state and fuel cycle codes	Data preparation, and steady state analysis design and operational physics support

TABLE 8.6/2

PWR DESIGN CODES

1. Total Plant Models

Code Name	Code Definition	Code Application
(a) SIBDYM UK CEGB/NNC PMSP	Sizewell 'B' total plant model includes reactor, pressuriser, steam generator(s), turbine(s) feed train, alternator and grid, boron thermal regeneration system, control system. Currently being converted to macro structure form	Operational dynamics control and intact circuit faults
(b) RETRAN USA NSSS only	General modular PWR model; includes reactor, pressuriser and steam generator Available through EPRI	Pressurised faults
(c) RELAP 4 MOD7 US	Modular 1D code. Homogenous fluid with bubble rise and liquid level tracking	LOCA studies
(d) RELAP 5 MOD1 US	Modular 1D code, 2 fluid, non equilibrium, 5 conservation equations	LOCA

(Continued)

TABLE 8.6/2 (Cont.)

(e) RELAP5 MOD2
US

As (d) but with full 6
conservation equation
representation

LOCA

(f) TRAC PD2
US

3D vessel representation 1D
pipework, 6 conservation
equations, non equilibrium

LOCA

2. Reactor Codes

(a) AKTS

3-dimensional 2 group reactor
steady state and kinetics
program. Provides a data
preparation route, 2 group to
1 group condensation and
reduction to 2-dimensional x,
y and 1-dimensional z models.

3,2 & 1-dimensional
steady state and
dynamic analysis e.g.
asymmetric rod faults
control system design
mass flow faults etc.

TABLE 8.6/3

CFR DESIGN CODES

1. Total Plant Models

Code Name	Code Definition	Code Application
FARESTAM UK CEGB/NNC PMSP	Station model, includes reactor, IHX, boiler turbine, control systems	Control and operational transients

2. Boilers

(a) NUMEL	Reference contraflow heat exchanger model has been configured to numerous plant specific once-through and drum boiler configurations	Detailed steady state and dynamic boiler analysis, static and hydrodynamic stability analysis
Also available in PMSP		
(b) DYMEL	Links to NUMEL for hydro-dynamic stability analysis	Hydrodynamic stability analysis

8.6.2   <u>Plant Analyser for the On-Line Determination of
Fault Recovery Strategies - Emergency Response
Plant Analyser</u>.

It is here presupposed that:

(a)   The plant analyser can communicate with the plant, probably by interrogating the plant data processing system
via an interface computer.

(b)   The information obtained from (a) defines the plant
state as defined by the process instrumentation, and
yields the plant dynamic boundary conditions such that
the plant model can be initialised and run for the
measured boundary conditions from an identified dynamic
or quasi-steady state initial condition.

(c)   The plant model is type specific, requires a detailed
data set for the given plant, and facilities for calcul-
ating core follow data, in the sense of ageing. Furthermore, faults may be super-imposed on normal operational
dynamics and hence a continuously operating front-end
processor is necessary to define the initial conditions
of certain integrators in the total plant model. This
is equivalent to the restart facility normally available
in large codes.

(d)   It may not be possible to deduce directly from the plant
instrumentation the cause of the fault. If this is the
case it is necessary to establish via the on-line analysis
facility the likely qualitative cause by logical operation
on the measured data. Given that the cause has been
correctly postulated it may still be extremely difficult
to calculate the quantitative cause. This is best illustrated by example. Suppose the fault was a medium
sized break in PWR primary circuit pipework. A prerequisite
to the calculation is the position in the circuit, the
hole size and discharge coefficient. The limit fault
analysis for licensing purposes assumes a double
guillotine fracture with non-interacting jets, whilst in
practice this is more likely to be a fracture which tears
into a complex geometry with an unknown discharge coefficient and area, both of which potentially vary in time.
The proposition is to the effect that it is possible to

iteratively determine the fault cause. Two points now
arise in the argument:

(i)   is such a calculation feasible? It is in
        principle, but it implies an even faster solution
        speed than that required to explore fault
        recovery strategies alone.

(ii)   is it necessary to effect such a calculation, or
        is it possible to develop the optimum fault
        recovery strategy without such detailed numerical
        analysis?

        Detailed consideration of this problem is required
        as the answer is probably dependent on the fault
        cause. It can be studied in the design plant
        analyser by using the analyser model to represent
        the plant fault, writing the plant measured
        variables to disc, and subsequently using the
        information to emulate fault dynamics of unknown
        origin.

(e) From the above one thing is clear, namely that betas of
     0.1 or less are imperative. It may even be necessary to
     ignore Design Quality Assurance and proceed with trivial
     models. It is equally clear that a specialised computer
     architecture, language and numerical techniques will be
     required for this type of analysis, and that this in turn
     will make the analyser plant type specific.

From the foregoing discussion it is clear that all aspects
of the Emergency Response Plant Analyser would have to be
emulated in the Nuclear Plant Design Analyser and proven
before detailed consideration could be given to plant implementation.

ACKNOWLEDGEMENTS

The authors would like to thank the Central Electricity
Generating Board of Great Britain for permission to publish
this paper. Numerous colleagues have made a contribution
to this modelling methodology over the years and this
contribution is duly acknowledged.

REFERENCES

[1] Gear, C.W. (1971) "Numerical initial value problems" Prentice-Hall series in Automatic Computation, New Jersey.

[2] Fowler, M.E. and Warten, R.M. (1967) "A numerical integration technique for ordinary differential equations with widely separated eigenvalues", IBM J.Res.Develop., 11, 537-543.

[3] Broyden, C.G. (1965) "A class of methods for solving non-linear simultaneous equations" Maths. Comp. 19, 577-593.

[4] Schubert, L.K. (1970) "Modification of a quasi - Newton method for non-linear equations with a sparse Jacobian" Maths. Comp 25, 27-30.

[5] Wilkinson, J.H. (1963) "Rounding errors in algebraic processes" NPL Notes on Applied Science No.32, HMSO, London

[6] Curtis, A.R. and Reid, J.K. (1971) "Fortran subroutines for the solution of sparse sets of linear equations", AERE Report R. 6844, HMSO, London.

[7] Porsching, Murphy and Refield "Stable Numerical Integration of Conservation Equations for Hydraulic Networks" Nucl. Sci. Eng. 43, 218-225 (February 1971).

[8] Wilkinson, J.H.(1965)"The Algebraic Eigenvalue Problem" Clarendon Press, Oxford.

MODELS AND SIMULATION IN NUCLEAR POWER

STATION DESIGN AND OPERATION

M.W. Jervis

Generation Development and Construction Division
Central Electricity Generating Board
Barnwood, Gloucester, GL4 7RS, U.K.

I. INTRODUCTION

Since the early days of nuclear power, models and simulation have been used extensively and they provide a strong integrating influence on the total systems design. Past and current practice of the Central Electricity Generating Board (CEGB), in the UK, provides a broad and well established background to illustrate typical usage of models and simulators, reviewed here.

Current techniques employ models and simulators in many areas and these include the following:

* full-scale and part-scale physical, spatial, models of plant as an aid to design of specific parts of the plant, particularly where there is congestion of pipes and cables and to investigate constructional techniques with photographic recording.

* full scale static mock-ups of control rooms, particularly control desks and panels, used for ergonomic development.

* full scale mock-ups with some dynamic displays.

* full scope replica control desks with displays and controls connected to an extensive plant model, forming a training simulator.

* large simulation facilities used for design purposes using mathematical models.

* nuclear plant analysers used in both design and operational phases, predictive facilities being used in an
emergency response mode.

* alarm analysis and Disturbance Analysis Systems and
predictive devices used as operator aids.

* small plant simulators used for closed loop testing of
control systems.

* calculation of the steady state, dynamic and transient
fault performance of the main electrical auxiliary
power and essential supplies systems.

* reliability evaluation of the main electrical auxiliary
power and essential supply systems.

* calculation and co-ordination of electrical auxiliary
system relay and fuse protection settings and characteristics.

* monitoring the integrity of essential systems.

These applications are reviewed with specific reference
to the common features in the various approaches and future
developments.

The characteristics of various types of simulators used
for training, design and as plant analysers are illustrated
in Figure 1 which also shows, very approximately, their areas
of function.

II. PHYSICAL FULL-SCALE ENGINEERING PLANT MODELS

As in many branches of engineering, full-size models
are used as mock-ups for checking the feasibility of arrangement of plant. In the nuclear engineering field such models
have been found particularly valuable in supplementing two
dimensional drawings of areas of plant where there is
congestion of plant components, cables and pipework. The
feasibility of construction can be established and operational
and maintenance procedures can be checked.

An extreme case [1] is a maintenance training facility
for BWRs that consists of disused plant which represents many
parts of the whole reactor plant. This full-scale facility is

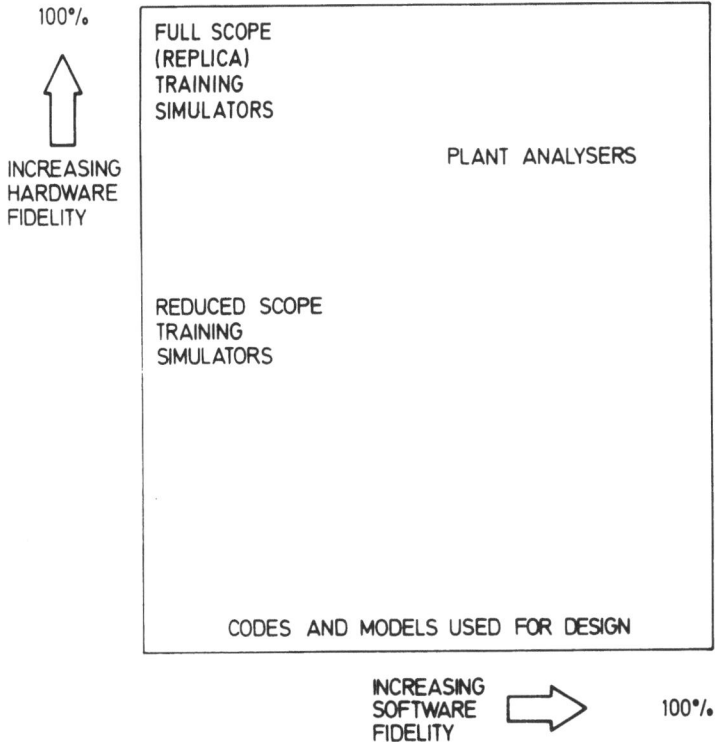

Figure 1
Roles of various simulators

used for training maintenance staff and those involved in refuelling.

III. PART-SCALE MODELS

Throughout the design process and particularly during
the detailing of the design, considerable attention has to
be given to constructability and use is made of a 1:75 construction model. The development of the station layout is
a major activity in the overall design of the station and
use is made of a 1:33 scale model. An example is shown in
Figure 2.

The detailed engineering layout can be established using
an Engineering Design Model typically constructed at 1:20 scale
with sufficient accuracy to enable it to be used for scaling
the dimensions for the construction of the plant. The model
is used in the final optimisation of the layout and details

Figure 2
Example of part scale plant mock up

the concrete and steel structures; the plant and equipment;
cable trays and ducting; pipework down to 15mm bore, with its
associated supports and restraints; heating and ventilating
ducting; and control and instrumentation equipment. This
information will be transferred to site with the aid of
photo-composites which provide dimensionally accurate photographs in plan and elevation of the plant. Larger scale
"engineering" models, typically 1:10 scale, are used to
investigate particular design problems which are dependent
on layout considerations. An important application of such
models is in the design of the safety related pipework.

The use of the model also allows 'walk-through techniques'
to be employed to investigate possible system interactions and

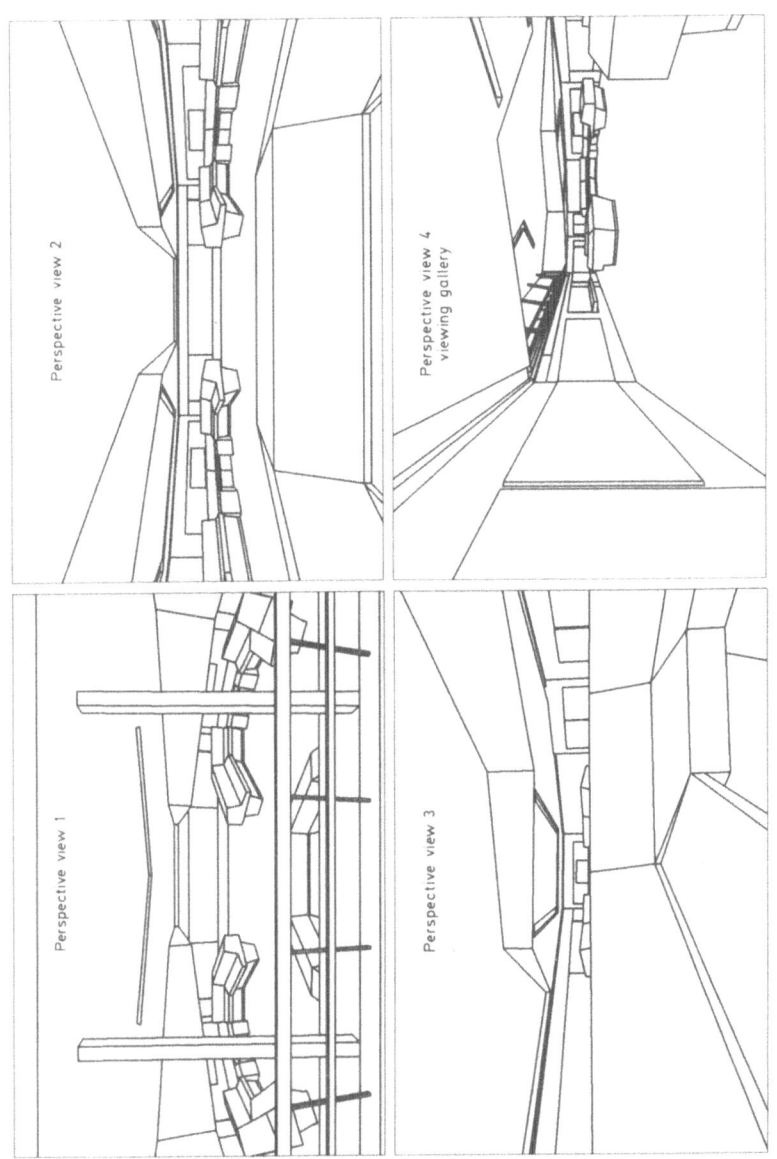

Figure 3.

Main Control Room layout produced by CAD
and showing views from four viewpoints

to ensure that the plant is optimised from a maintenance viewpoint and particularly with regard to reducing exposure of
staff to radiation. A system of comprehensive reviews can be
carried out by multidisciplinary teams when the relevant parts
of the layout are 'frozen'.

Reduced-scale models have also been used during the
design of control rooms, an introscope being used to view the
model from various angles to check lines of sight. This
method is not very effective and has been superseded by the
use of Computer Aided Design (CAD) techniques that provide
perspective views. An example is given in Figure 3 which
shows Heysham II Advanced Gas Cooled Reactor (AGR) power
station control room.

IV. FULL-SCALE STATIC CONTROL ROOM MOCK-UPS

For many years the development of the man-machine interface has been assisted by the use of static full-scale mock-
ups of all or part of the control room facilities as
illustrated in Figure 4.

Such static display has been very useful in developing
and checking such factors as:

* anthropometric matters e.g. reach, knee room, viewing
  distances, lines of sight etc.

* relating the results of task analysis to the organising
  of indications and controls to maintain proper association
  in the form of functional groups.

* developing and validating operating procedures using the
  mock-up in 'walk-throughs' and 'talk-throughs'.

The use of such mock-up techniques can be applied to any
form of control desk but it is particularly effective when
the desk is of the modular type. In the CEGB implementation
[2] the desk surface is divided into 72mm squares. These
conform to DIN standards and the desk supporting framework
accommodates these either as single 72mm square modules or
multiples, 72 x 144mm being common. The modules are fitted
with a range of indicators, lamps, alarm windows, push
buttons etc. and have plug and socket connections at the back.
If maintenance is required, the modules can be withdrawn from
the front without disturbing other modules and the defective
module can be removed quickly and replaced by another one.

Figure 4

Example of simplest form of control room mock up.
The post trip mimic display is directly behind
the desk Section.

The system has the advantage of great flexibility in
that the positions of the modules can be changed at a late
stage of construction, and even during operation, e.g. if
operational modes or procedures are changed and corresponding
layout changes are desirable. The number of different modules
is kept relatively small and additions are included only
after careful consideration and justification.

The use of this system facilitates a formalised approach
to design and documentation. At the design stage the appearance
of each module is held in the CAD model and can be called up
and rapidly assembled to produce a complete layout drawing.
The positions are identified by their rectilinear co-ordinates
on the desk or panel surface and are entered into a comprehensive schedule that provides a complete computerised
information base for control and instrumentation hardware

for the whole station. This is applied 'cradle to grave'
from the initial design stages through to final design
including cable information, ordering, progressing maintenance
and ultimate replacement. For example, existing modules may
need to be replaced because, say, moving coil meters may be
advantageously replaced by plasma or solid state displays.

With this approach the information about the control
desk and its module population can be readily transferred to
the construction of both the static mock-up and to a replica
training simulator if one is to be provided. Any modifications
agreed during the mock-up stage can be readily transferred to
the actual desk.

The overall concept embodies a formalised approach to
plant nomenclature and labelling which is documented in the
scheduling system and this forms an essential part of the
design information base for the power station. The documentation is also used during the evolution of the control system
with the relevant manual controls and indications being
readily available for transfer to the control system diagrams.

The limitations implicit in static mock-ups are mainly:

* lack of a time element, and absence of movement of
  indicators, controls, lighting of lamps etc.

* lack of a consequent effect after operation of controls,
  giving feedback.

V. FULL-SCALE CONTROL ROOM MOCK-UPS
WITH DYNAMIC DISPLAYS

To avoid the limitations of static mock-ups, they have
been enhanced by making some of the displays 'live', driving
them by electronic circuitry to give some simulation of the
dynamic behaviour of the indications. The circuitry is
usually of the 'playback' pre-programmed sequential type with
little or no interaction, as illustrated in Figure 5.
Typically, a small personal computer is adequate for this
purpose.

The limitations of simply dynamic displays are mainly:

* representations of only a limited part of the total system

* reliance on a pre-programmed rather than interactive operatio

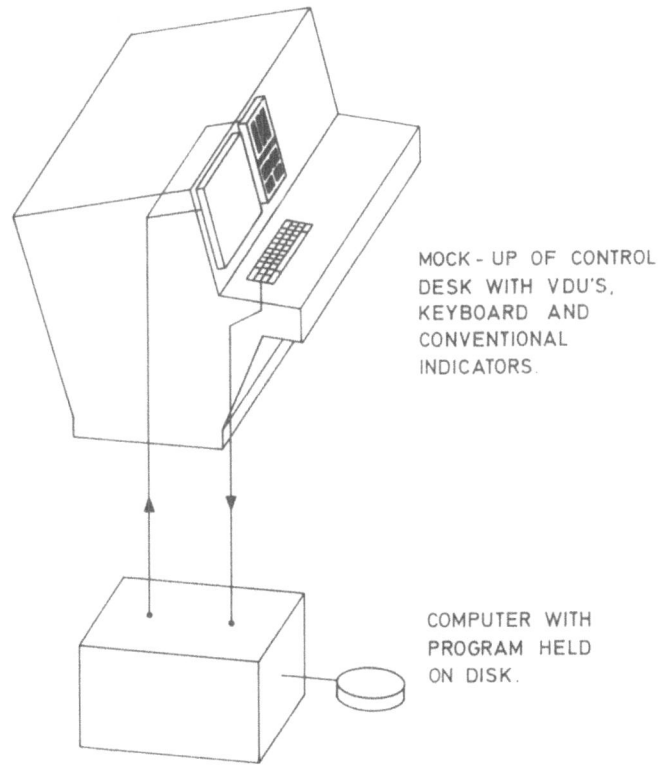

MOCK - UP OF CONTROL
DESK WITH VDU'S,
KEYBOARD AND
CONVENTIONAL
INDICATORS.

COMPUTER WITH
PROGRAM HELD
ON DISK.

Figure 5
Scheme for program driven displays

Nevertheless, they have been found to be valuable as
design aids and as demonstrations to operators and regulatory
bodies. Two examples will be described.

Post trip monitoring panel

On the AGRs, a reactor trip causes the post trip
cooling system to be activated and various automatic actions
take place to ensure that sufficient reactor cooling is
available to remove the decay heat from the reactor. It is
helpful for the operator to have a clear understanding of
the progress of the sequences and a mimic diagram display is
provided in the control room [3] as shown in Figure 4.

In the early stages of its development, this display
was designed on the basis of a static mock-up but there is

great advantage in being made 'live' and operate in real time
corresponding to the various reactor trip scenarios that can
occur. An actual prototype of the display was produced and
was driven, through interface equipment from a BBC Model B
personal computer, the software being written in BASIC, with
only a few man months of effort.

This simple unit has enabled the design to be checked
and refined. Being portable, it is easy to take it to the
power station site and it has proved invaluable in demonstrating to station operating staff and other concerned
parties, how the system will work. The unit will be incorporated later into the full scope Heysham II training simulator.

On other stations, for example, Hunterston B and Hinkley
Point B, post trip simulators of this type have been provided
at the power stations to provide refresher training to
supplement that provided at the 'central' training centre.

Development and Demonstration of Data Display and Alarm Systems

During the development phase of alarm systems of the AGRs
and PWRs a system was set up by CEGB using DEC LSI11 computers
and associated disc, printers, keyboard and VDU display drives
and a variety of VDU monitors. The software ran under RSX-11
with programs written in CORAL 66 and PDL (Plant Display
Language).

The displays were selected by a keyboard and enabled up
to three colour VDUs to be used at the same time.

The equipment was used for three basic purposes:

* assessment of hardware, for example, a comparison of
  triad and in-line gun configurations and need for long
  persistance phosphors to reduce flicker

* ergonomic studies on optimum formation of alpha-numeric
  characters, number of lines per page, character aspect
  ratio, space between lines etc., also keyboard facilities,
  protocols etc.

* demonstrations of plant incidents in real time

As an example of a demonstration, a sequence of alarms
with their timings was obtained from the various sources that

reported the TMI -2 incident and these were listed in time
order with the intervals between their occurence. They were
loaded into the computer system and the program then presented
them on the log and on the VDUs at the correct times. The
appropriate data displays were also updated with data obtained
from the published literature and those showed the variations
of pressure, levels and temperature during the times the alarms
were active.

The system provided a system for handling the alarms e.g.
accept and reset and had an 'all alarms' list and a 'plant
area' alarms list [2,3]. However, it was not truly interactive in that there were no facilities to affect simulated
plant operation.

The demonstration made it very clear that the operators
had severe problems of information overload and it enabled
the designers to develop strategies for data display and
alarm handling to ameliorate these problems. Some of these
have been incorporated in future designs [2,3] and these
improvements are being tested to check the advance that has
been made.

A basic problem of this approach is the availability of
a sequence of alarms and data variations that represent plant
incident. A nuclear plant analyser could in principle
produce such scenarios, the range of scenarios being generated
by a superimposition of fault situations [4], generated by a ·
'simulator driver' discussed in Section XII.

VI. FULL-SCOPE TRAINING SIMULATORS
AND ENGINEERING SIMULATORS

6.1 General

The use of simulators for the training of nuclear power
station operators is now well established. Most countries
that have a nuclear power programme have training centres
and these employ simulators.

As an example, in the UK, the Nuclear Power Training
Centre, operated by the CEGB and located at Oldbury, Avon,
has a generic Magnox simulator and three AGR plant specific
replica simulators with a further one for Heysham II AGR
under construction [5,35] . The South of Scotland Electricity
Board in Scotland has a comprehensive simulator for its
Hunterston B AGR [6,7].

The heart of any training simulator is the modelling software [8]. The hardware of such systems typically comprises:

* a large computer system, on which the plant models are run in real time, discussed further in Section XIV.

* a replica control desk and with panels complete with indicators, controls, and VDUs with a computer system driving the control indicators and accepting manual control inputs.

* replica data processing system hardware and software providing the alarm and display facilities including driving the desk VDUs.

* a set of Tutors' facilities.

The design and construction of the replica control desk and panels bears a very close relation to the actual ones used in the power station. The design is developed using the static mock-up described in Section IV and dynamic mock-ups of the type described in Section V. In this connection, a strong unifying influence is exerted by the adoption of the modular approach to desk construction.

6.2 UK AGR Training Simulators

In the AGR training simulators, the modelling codes that are produced for design purposes are also used as a basis for the modelling software in the training simulators [8].

In the case of the CEGB simulators installed at the Nuclear Plant Training Centre [5,35], the models are run in the CEGB's central IBM 3801 mainframe computers in London. The computers communicate with the simulators themselves at Oldbury, Avon, some 200km away, by a high speed data link.

The hardware for the Hunterston 'B' AGR simulator [6,7] consists of 52 Marconi GRADUATE computers arranged as parallel processors. Generally, these are configured to take advantage of the parallel nature of the plant, this contributing to the high speed real time solution. The size of the model is indicated in Table 1.

The simulator is equipped with replica desk and panels and a Tutors' facility. It is the latter coupled with the

TABLE 1

ESTIMATED NUMBERS OF DIFFERENTIAL, ALGEBRAIC AND BOOLEAN EQUATIONS
FOR THE HUNTERSTON 'B' AND TORNESS SIMULATORS

Plant Module	Hunterston 'B'			Torness		
	Diff. Eqns.	Alg. Eqns.	Boolean Eqns.	Diff. Eqns.	Alg. Eqns.	Boolean Eqns.
Reactor - xy point representation	3359	2720	-	7636	6640	-
- 9 plane axial	353	180	-	353	180	-
Rod Controllers	20	730	-	225	765	-
Gag Flow	-	308	-	-	332	-
Gas Circulator through boilers etc.	20	308	-	20	333	-
Circulators	24	96	-	24	96	-
Electrical System	20	1000	-	20	1000	-
Boilers	1170	2700	-	1690	3900	-
Turbine	120	550	-	120	550	-
Control Systems - excluding rods	33	200	-	33	200	-
Protection system & boiler trip logic			5000			5000
Feed pumps - BFPT & 2 x SU/SB	10	50	-	10	50	-
Gas clean up & chemical monitoring	-	-	-	20	50	-
TOTALS	5129	8842	5000	10,151	14,096	5000

Diff - differential
Alg - algebraic

high degree of real plant emulation and rigorous data route
which makes this device capable of being used as a plant
analyser, and as a design tool for proposed modifications to
the existing plant.

6.2.2 Hunterston 'B' Simulator [6,7]

The South of Scotland Electricity Board and Marconi
Instruments have developed a simulation of the Hunterston'B'
Advanced Gas Cooled Reactor Station, the original intent being
for use as a plant simulator, but with the inherent capability
for use as an analyser. The real time solution has been
obtained by a unique tailoring of hardware and software,
arrived at by careful consideration of the characteristics of
the equations. This approach was necessary to achieve the
fundamental objective of the simulator which was to use the
same models as in the original plant design codes and to give
a realistic representation of the asymmetry implicit in the
plant.

The simulator software was developed from design models
that had been developed by the Central Eelectricity Generating
Board for studies on the similar Hinkley Point 'B' Station.

6.3 Engineering and Conceptual, Principles, Simulators [9,26]

Full-scope, replica, simulators are needed for effective
operator training. However, small simulators with a minimal
operator interface and relatively small models and computers
can be used as 'engineering' 'conceptual' or 'principles'
simulators for the familiarisation of engineering and operational staff with the basic plant dynamics. These have
proved to have an important role complementing the full scope
training simulators [9,26].

VII. LARGE SIMULATION FACILITIES USED FOR DESIGN

7.1 Plant Kinetics

Since the very early days of nuclear power, simulation
facilities have been used in the design of nuclear plant in
connection with fault studies and control system design [8].
The early systems used operational amplifiers in analogue
simulators and these were displaced by hybrid systems and
currently digital simulation is almost universally used. Another
example is discussed in the following section.

7.2 Electrical Auxiliary System Design [10-12]

The analysis of main electrical auxiliary and essential
supply power systems is now undertaken with digital computer
systems, which combines a suite of interactive power system
programs with full network drawing and graphics facilities.

The software packages essentially incorporate comprehensive load flow, short circuit fault, dynamic and transient
fault stability performance facilities.

Electrical power systems are entered on the computer by
drawing the system network using standard symbols on a graphics
terminal screen. Tabulated data pages are automatically generated from the network for the input of the appropriate
system and plant parameters including the appropriate Generator
Governor and Automatic Voltage Regulator (AVR) control models.
Load flow and fault study key results can be displayed
directly on the network diagrams, while for dynamic and
stability studies the voltage, slip, angles, etc. can be
plotted graphically against time.

The predictive performance of electrical power systems
under normal and abnormal conditions is essential in the
design phase for the optimisation of system and plant
parameters and the subsequent planning and operation needs
of a power station.

A graphic interactive reliability analysis computer
program with an identical electrical power system network
drawing facility is also used for evaluating at each system
voltage busbar level the three main reliability indices
namely, the expected failure rate, average outage duration
and average annual outage time. The digital software program
incorporates realistic failure modes including passive and
active failures up to third order, stuck breaker conditions
and the effect of maintenance and realistic restoration modes
such as repair and replacement, switching, alternative
sources of supply, standby plant and common mode failures.

In order to achieve the optimum discrimination in operation
of the electrical power system protection relays and fuses an
interactive computer program is used to calculate relay settings
and produce a graphical display of the grading curves for the
relays and fuses. The program determines the appropriate
relay settings for a three-phase fault and checks the validity
of the settings under various transient fault conditions.

VIII. NUCLEAR PLANT ANALYSERS

8.1 Background

It has been suggested that the use of 'plant analysers'
or 'design simulators' would be a valuable tool to assist in
the design and/or operational phases of nuclear power
plants [13-16]. In this paper the term 'nuclear plant
analyser' ("NPA") will be used. NPAs can be described as
devices which comprise a digital computer complex which can
run comprehensive models of the plant concerned and display
the results via a suitable operator input and output interface as illustrated in Figure 6. They can be used in the
design stage to assist the safety analysis of plants, in the
operational phase, particularly in severe fault conditions,
and as an aid in training programmes.

While such facilities have similarities with training
simulators there are major differences as illustrated in
Figure 1. Full-scope training simulators [5-7,35] are
provided with an operator interface that is a close replica
of the actual plant Main Control Room (MCR). The digital
simulation runs models with sufficient fidelity to permit
adequate training for the defined set of conditions required
by the training programme. The models usually run in real
time.

Plant analysers have detailed models that can be used
to assist in the design or operational phase to cover a wide
variety of situations. It is not necessary to provide a
replicate MCR and the NPA does not necessarily operate in
real time, (Figure 1).

8.2 Benefits from the use of Nuclear Plant Analysers

8.2.1 Summary

The situations in which NPAs provide benefits include:

* the design stage of the power station

* during operation of the power station

* in training programmes

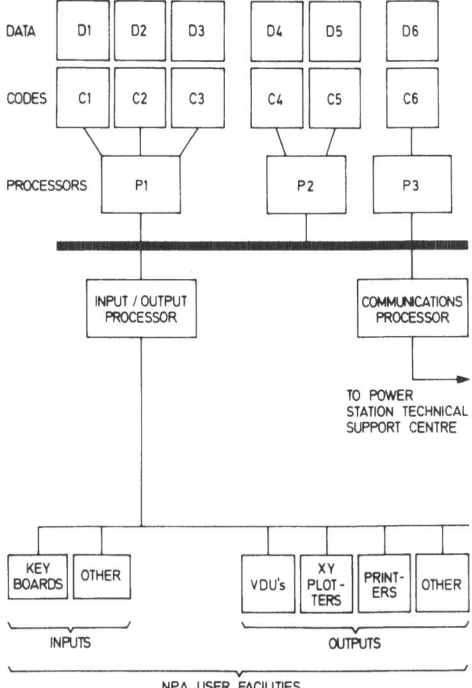

Figure 6

Basic Nuclear plant analyser configuration

8.2.2 Benefits of Use During Design

The benefits that can be expected include the following:

* ability to perform safety analysis on a multi-user
  basis in a more effective, faster and cheaper way than
  is done at present.

* assistance in code validation against operating data.

* development and validation of Station Operating
  Instructions and Operating Rules.

* cumulative damage analysis e.g. creep, fatigue, corrosion.

* exploration of plant operating conditions inside and
  outside the design base events and formulation of operator
  guidelines and special displays to cover these cases.
  These conditions include the superposition of faults on
  normal operational dynamic situations.

* allowing the analyst to assimilate multi-parameter
  variations in an efficient way which will lead to a
  better understanding of plant behaviour. In addition
  this rapid interactive ability allows identification of
  fruitful lines of investigation and elaborate sensitivity
  analysis.

* familiarise design staff with the transients that plant
  will or may experience during operation.

* development of VDU displays and other parts of the man-
  machine interface [15,17].

8.2.3   Benefits of Use During Plant Operation

8.2.3.1   General

The primary benefits that have been identified
for NPAs are:

*   perform system analysis within a short time of a
        plant emergency, to diagnose the state of the
        stricken plant, and formulate recovery actions.

*   update safety analyses as a result of plant mod-
        ifications or the effects of plant ageing e.g.
        fuel burn-up distributions.

*   investigate operational strategies e.g. load
        following manoeuvres, to ensure maintenance of
        fuel safety limits etc. These can be described
        as "what if?" studies.

*   maximise power output and assisting in rapid start
        up after plant shutdowns.

In addition, NPAs may provide assistance in the following
areas:

(a)   The analyser appears to be the best and most
        efficient system to utilise the engineering codes:
        it permits visualisation of the results in various
        forms, it permits interaction and may be made
        "user friendly" by exploiting colour graphics
        VDUs.

(b)   By proper interface between the various codes, the
        analyser can extend its simulation to a variety
        of accident situations involving many parts of a
        complete station making use of different computat-
        ional methods. However, considerable development
        is required to make this a reality.

(c)   In the formulation and updating of Operating
        Instructions and operator's manuals.

(d)   In the licencing of plants, the re-licencing after
        repairs or system modifications. The analyser
        could provide the means to substantially reduce
        delays in licencing.

(e)   In the testing of single components, to test the
        control systems, the communication and measurement
        systems, the diagnostic systems.

(f)   As a training simulator whenever the physical
        replica of the control room is not essential, for
        example, in the rehearsal of transients by senior
        operators or in the training of 'analysts',
        Technical Support Centre engineers and similar
        expert personnel; in emergency exercises.

        The analysers, as presently conceived, do not
include any diagnostic feature, and it is not clear by
how much they might be of assistance to operators during
an emergency, as it is sometimes claimed they would be,
and this requires careful consideration.

        This function can be generalised as providing in-
depth assistance to the Main Control Room (MCR) operators.
It is important to consider the role of the Technical
Support Centre (TSC) because in many plants this will be
the route through which the MCR operators receive this
assistance.

8.2.3.2 <u>Role of the Technical Support Centre (TSC)</u>

Although dedicated "per plant" NPAs may be justified, it appears likely that the cost of a NPA will be
so high that it would be shared between several power
stations. However, such sharing raises significant
technical and commercial problems.

A likely way of using such an NPA would be to
assist the station operational staff by carrying out the
required dynamic analysis and transmitting the output to
the station. Each station will probably be provided
with a Technical Support Centre (TSC) [18], as shown in
Figure 7. The facilities provided in the TSC are likely
to include means by which:

(a)   information on the operational state of the power
         plant is transmitted from the station to the
         analyser

(b)   the results of the analyses and consequential
         advice are transmitted to the MCR operators.

Under these circumstances, the ease of transfer of
information and advice would be maximised by arranging
for the NPA operator interface to resemble fairly closely
the TSC operator interface. Since the means of presen-
tation and input are available for the plant analyser,
there would be no difficulty in providing similar pres-
entation in the TSC but some co-ordination would be
desirable.

8.2.3.3 <u>On-Line Diagnosis and Prediction</u>

Although it is likely that NPAs will first be
established as a means of providing operational guidance
via the TSC, it is possible that later developments will
enable their use as short-term decision aids within the
control room. In principle, a plant analyser could be
directly coupled to the plant and used to:

i)   provide estimates of those aspects of the plant
         state which are not measured directly ("state
         estimation", "virtual instrumentation" or "analytic
         redundancy" techniques [19,20].

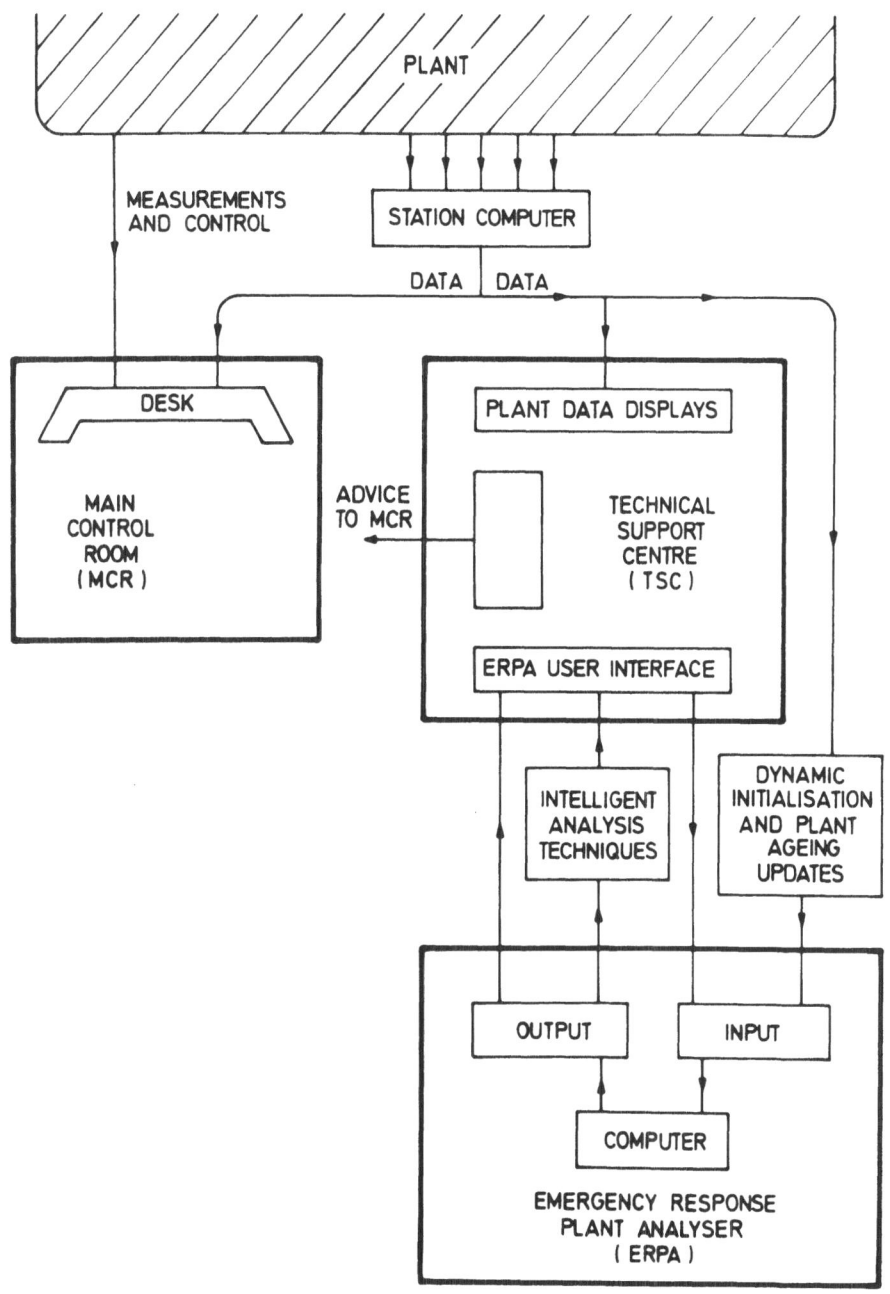

Figure 7
Nuclear plant analyser with facilities
for emergency response role

ii)   provide indications of plant malfunction, detected
        as discrepancies between NPA behaviour and
        measured plant behaviour.

iii)   predict the future prospect of an undesirable
        plant state.

iv)   evaluate the effectiveness of alternative oper-
        ational strategies.

A NPA provided for any of these purposes would be
required to capture and assimilate plant data reliably:
a link is shown in Figure 4. The monitoring and functions
(i) (ii) would require a facility capable of real-time
simulation, while the predictive functions (iii) and (iv)
would require a simulation speed at least an order of
magnitude higher. Furthermore, the plant data would have
to be correct, otherwise the prediction could be mis-
leading. For these reasons, the provision of NPAs for
on-line diagnostic and predictive support for the plant
operator is seen as a longer-term advanced application
requiring significant development.

However, as described in Section IX, a system
having some of the basic components is already operational.

Transmitter Validation

The availability of an NPA, receiving on-line data and
plant states via the unit computer system, provides the opportunity to use the analyser in a diagnostic and audit role.

For example, the analyser can provide a modelling facility
with a model, the inputs of which include the outputs of plant
transmitters and demands to actuators that control the plant.
The output of the model then indicates the expected states of
the equipment and plant conditions and these can be compared
with the actual, measured, ones. Assuming a good model, any
differences could be detected and used to identify faulty
equipment.

In the case of actuators, a demand should cause an effect
on the plant that can be modelled e.g. a demand to open a
valve should cause a predicted increased flow. If this does
not occur, the actuator performance is suspect, e.g. travel
time is longer than that specified.

For transmitters, their outputs should be consistent with the plant conditions in the model and the outputs of other transmitters measuring related plant conditions. By checking for inconsistencies between the model and transmitter outputs, suspect transmitters can be identified.

Such a system could be used in a diagnostic role, indicating faulty equipment and the need for maintenance of transmitters, actuators and other equipment.

8.2.4   Benefits to Training Programmes

8.2.4.1   Use in Conjunction with Training Simulators

Most countries that operate nuclear power stations have training simulators [5-7] and NPAs could be used to assist in developing the training programme and its equipment.

A design simulator would be valuable in assisting in the development of the specification of such training simulators and their design. Some of the models run on the NPA may well be used as a basis for the training simulator models.

The NPA could be used to investigate situations not currently covered by the training simulator models. These tend to be limited by economic considerations, or the adoption of a policy of continuous improvement and development which takes advantage of the increases in computing speed of newly available hardware.

Having explored a situation on the NPA, the results ("scenario") could be transferred to the training simulator and then used for training purposes. In its simplest form, such scenarios could be recorded and played back through the training simulator displays, in a non-interactive mode, for instructional purposes.

If exploration on the NPA reveals situations that are important to training, the analyser could be used to identify the situations that have to be modelled for interactive training purposes. Furthermore, operator errors revealed during use of the training simulator could be investigated on the NPA, particularly if the action

takes the plant outside the capability of the training
simulator.

8.2.4.2   Training of Instructors

An NPA would provide a means of assisting operator
training centre instructors to appreciate plant situations
beyond the scope of replica training simulators. This
would enable instructors to reach the superior level of
knowledge of plant operations necessary for the
effective instruction of operators.

8.2.4.3   Advanced Training of Technical Support
Centre Personnel

TSC personnel are expected to advise station
operating staff when the TSC is used in an emergency
situation [18]. The use of a plant analyser in the
training of TSC personnel would be very helpful in
making them aware of a wide range of possible plant
situations not simulated on the replica operator
training simulators.

8.3   Use of NPA in ergonomic design

A plant analyser which contains a fairly comprehensive
model of the plant gives the opportunity for providing a
much more effective assistance to ergonomic design.

When used in the arrangement shown in Figure 5, the
system can be used for all the investigations listed in
Section IV with the additional features of realistic:

* interactive effects between controls and indication,
  alarms, etc.

* rates of movement of indicators and changes in VDU
  displays, printouts etc.

* rates of occurrence of alarms

These features enable the operator's reactions and
capability to be properly assessed and are considered
essential for the effective development of the ergonomic
development of the man-machine interface [15,17].

IX. PREDICTIVE DEVICES USED AS OPERATOR AIDS

If a model of the plant is fed with data measured on-line from the plant, the output of the model can be used as a valuable operator aid [21].

In one system now operating in a coal fired plant [22] an on-line model is used in an analytic mode to assist the operator in interpreting the data collected from the plant. By the use of data validation techniques and a user-friendly interface the system gives credible, relevant, convenient, immediate and unambiguous assistance to supplement the raw data displayed through the conventional data processing function.

The system also enables 'what if' questions to be posed and answered. These include:

* systematically investigating the effects of changing the principal controllable plant parameters.

* exploring the effect of changing items or groups of controllable data.

* exploring the effect of changing items in the model data.

X. HEYSHAM II POWER STATION FAULT LEVEL
MONITORING AND INDICATION EQUIPMENT

On-line models of the electrical networks of power stations can be used in operator aids. One example is a monitoring and indication system that is to be applied to the Electrical Auxiliary Supply System for the Heysham II AGR power station which provides all the electrical power for the operation of auxiliaries within the station.

The Electrical Auxiliary Supply System (EASS) is remotely controlled by Operators in the Central Control Room (CCR), who must configure the system in a manner than ensures that the prospective fault current, seen by switchgear, never exceeds the switchgear's fault level capability.

In the past this has been accomplished by installing a hard-wired electro-mechanical relay interlock scheme, which involved directly wiring together various combinations of switchgear auxiliary contacts into the closing circuits of switchgear. Thus a particular item of switchgear could only be operated if the state of auxiliary contacts of other switchgear were in a configuration that allowed completion of the operating circuit in question.

Early in the design of the Heysham II EASS it was apparent that the design and implementation of such a hard-wired interlocking scheme described above would have presented enormous difficulties, particularly in the essential electrical system. Many simplifying assumptions would have had to have been made, which would have resulted in an inflexible system, preventing many desirable switchgear configurations, particularly those encountered under abnormal operating conditions.

As an alternative, it was decided to develop a computer based fault level monitoring and indication system to model a selected part of EASS, in real-time, and determine from the state of the circuit breakers in the system the prospective fault levels. This Fault Level Monitoring and Indication Equipment (FLMIE) will then be used to determine the circuit breakers which cannot be closed without exceeding the switchgear fault level capability. A monitor in the CCR can display the connections and switchgear for any of switchboards included in the system. The display identifies any switchgear which cannot be safely closed at any instant in time and gives the potential increase in fault level if they were to be closed. In addition, each switchgear control switch in the CCR will be provided with a red lamp, which will be illuminated if it is unsafe to close the circuit.

The information provided by the computer will be used as an advisory aid by the operator. Although a computerised interlocking system has been installed at the Drax coal-fired power station, it was decided not to provide full interlocking at the Heysham II nuclear power station. This decision was taken so as not to inhibit the operators from re-establishing vital electrical supplies under controlled emergency conditions to ensure reactor safety.

The fault level calculation and graphical display software for the system is being written by the University of Manchester Institute of Science and Technology (UMIST). It is based on the UMIST "IPSA" package [23] with modifications to enable the software to run in real-time; provide colour displays; acceptable for use in a CCR; and switchgear safety status information, by the lamp states.

The response time of the IPSA fault level calculation algorithm has been improved by obtaining the impedance matrix of networks using diakoptical methods [23,24]. (Diakoptical, from the Greek dia=through, and kopto=to tear, is the name given by Gabriel Kron to his method of subdividing large physical problems to enable their solution to be simplified. For descriptions of Kron's work see reference [24]).

The software has been specified to be flexible and allow circuits and switchboards to be added and removed, should the Heysham II EASS change in the future. It also follows that the software could be used on future or existing power stations by modifying the network and data in the computer model. In practice this could be achieved by an Operator drawing a new electrical network on a suitable graphic terminal, using predefined elements within the software drawing package.

The software is being installed in a 32-bit computer by the Solartron Systems Group which is also writing software for the Operator interfaces and all data acquisition. A schematic of the hardware configuration is shown in Figure 8.

Analysis of the hardware and software has shown that, with a fault coverage of 80 percent, a reliability of one fault in 10 years can be achieved between complete FLMIE system failures, which meets the specified requirement. The FLMIE at Heysham II is due to be installed in 1986.

XI. CEGB ESSENTIAL SYSTEM INTEGRITY MONITOR AND EMERGENCY COOLING SIMULATOR

11.1 Essential Systems Integrity Monitor [25,26]

The safe operation of nuclear power stations requires that systems performing essential functions are available with an appropriate level of integrity. One example is the system provided for post-trip cooling to remove heat following a reactor shutdown.

Such systems are complex and employ redundancy to provide the essential functions with high reliability in the presence of faults or plant taken out of service for maintenance. The minimum level of plant redundancy of such systems can be defined to the operational staff in the form of written instructions.

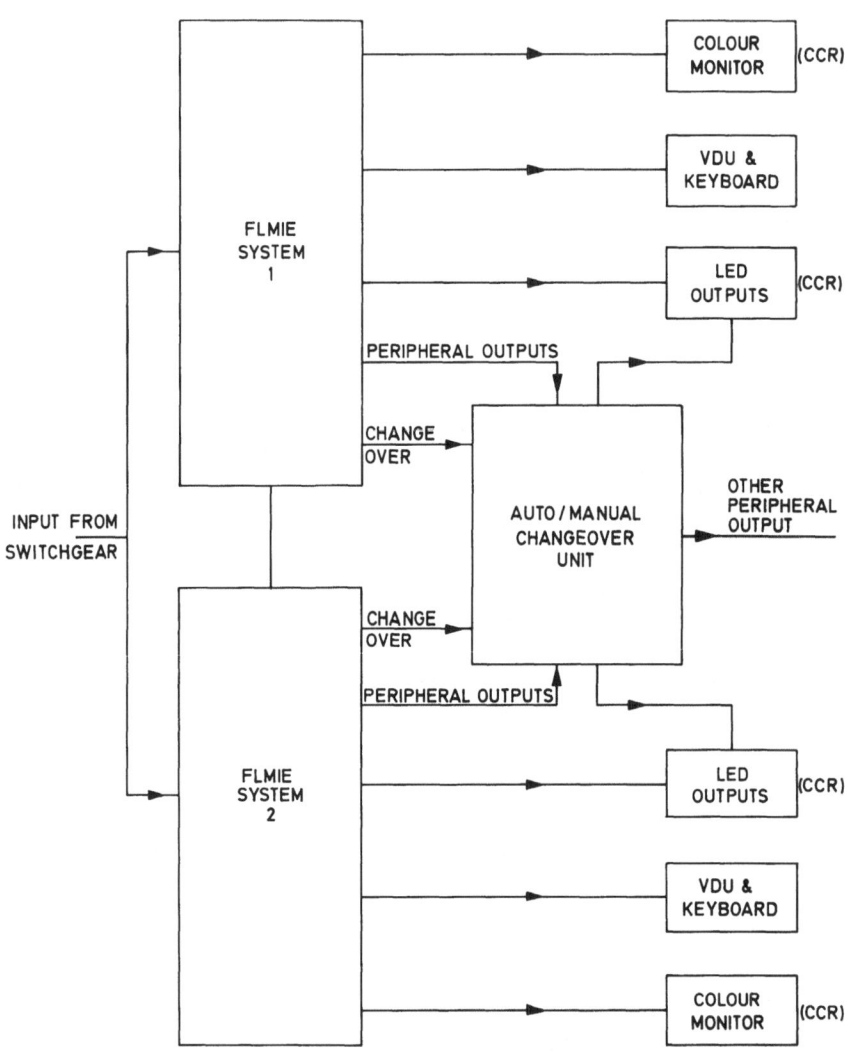

Figure 8
Hardware configuration of Fault Level
Monitoring and Indication Equipment

The systems and requirements are complex and the instructions tend to interpret the requirements for minimum levels of plant redundancy in ways that are pessimistic.

A computerised aid described by HORNE [25,36] is being developed for CEGB AGR power stations to provide the operators with an indication of the implications of plant outages on the integrity of the essential post trip functions. The device can also indicate the capabilities of the essential systems to tolerate further plant outages and assess the overall reliabilities under these conditions. This assessment can be used to plan maintenance strategies.

The system operates on the basis of fault tree modelling. These are logical diagrams constructed to represent all the combinations of failures of individual plant items which would lead to overall system function failure. This model is run in a digital computer with appropriate keyboard input, VDU display and printout facilities.

11.2 Emergency Cooling Simulator [26]

CEGB has developed a small device to simulate fault conditions, illustrate the design intent of autostart of the emergency cooling motors and pumps and act as a training aid so that in the event of failures of the cooling system, correct remedial action will be taken [26].

XII. EQUIPMENT, SYSTEMS AND HUMAN RELIABILITY

The techniques of equipment and system reliability analysis and prediction have become an essential feature of nuclear power engineering. These rely on the availability of failure rate data and reliability models [27].

It is recognised that the human operator plays an essential part in nuclear power station operation and so the question of quantification of human reliability arises. One approach is to develop techniques for modelling potential human errors in proceduralised and complex situations [28] for example based on the model developed by Rasmussen [29,30] shown in a simplified form in Figure 9.

Since the operator is an integral part of a man-machine system the human model has to be considered in close relation to the remainder of the system [28,29].

TYPE OF BEHAVIOUR   DECISION MAKING ELEMENTS

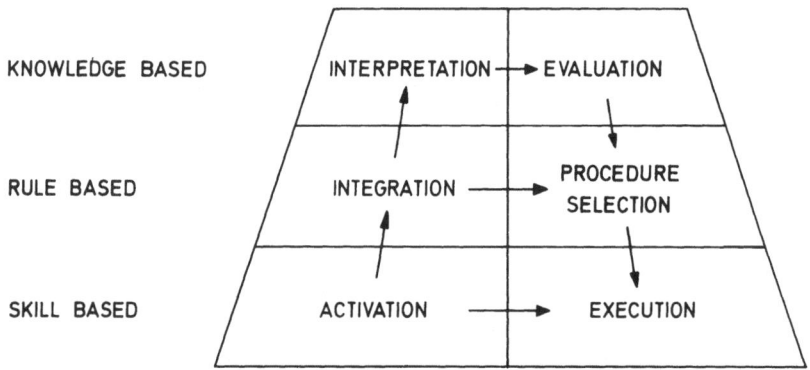

Figure 9
Simplified Rasmussen Model

It has been proposed by Cacciabue et al, [37] that a
model of the operator can usefully be coupled to the output
of a plant model so that the response of the total plant-
operator system can be investigated.

When stimulated by an automatic 'Simulator Driver' of the
type discussed in Section V, a large number of fault situations
can be investigated on an automated basis and the performance
of the total plant-operator system to be evaluated and
checked for acceptability.

This concept requires the availability of:

* The simulator driver, which must be capable of generating
  sequences of events, either systematically or randomly;

* The plant behaviour model, which accounts for the actual
  rector as well as for all the interconnected safety
  systems;

* The modelling of the operator's interventions developed
  in a fully interactive fashion with the other systems
  simulated by a Response System Analyser[37].

XIII. TESTING OF AUTOMATIC CONTROL LOOPS

The hardware and software that provide the continuous 'modulating' auto control of the various plant systems in a power station take the form of either

* analogue systems using 2- or 3-term controllers with some term adaptation and logic switching system

* direct digital control in modern schemes employing distributed microprocessors

* analogue controllers with some modification of set points by a digital computer

Having assembled such a system it can be tested 'open loop' by feeding it with dummy inputs to represent the output of plant transmitters and the plant states e.g. voltage to represent thermocouple outputs on on/off switches to represent switched signals. The output signals can then be measured and compared with the expected output which is related to the transfer function of the control system.

Such open loop testing has limitations, mainly associated with the interpretation of the outputs.

A more effective approach is to connect the control system to a simulator that provides a model of the plant and then check its performance. The arrangement is shown in Figure 10. In this way, many of the aspects of the control system can be checked and it provides the opportunity to examine the effects of changing the control loop terms such as gain, integral and derivative action, deadbands etc. With additional complexity, stability margins can be checked.

While still in this development stage the simulator and control system can be tested in conjunction with an appropriate subset of the total man machine interface. Typically this comprises some VDUs and a set of modules of the type described in Section IV on a basic frame, these including indicators, lamps and switches that are associated with the control loops concerned. Thus the responses of these devices can be checked well in advance of them being required at site.

When the control system has been delivered to site and connected to the actual control desk, the simulator can again

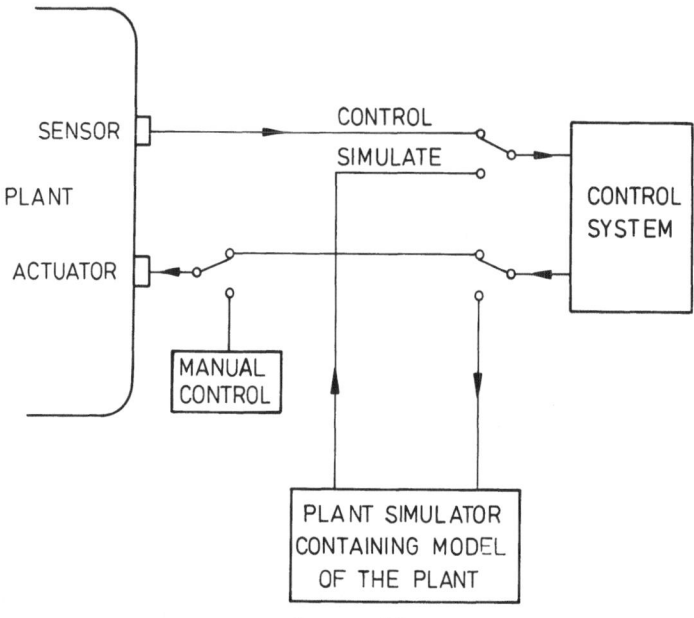

Figure 10
Closed loop testing using plant simulator

be used. At this stage it can be used to check for cabling
and termination errors so that they can be corrected early
in the programme. CEGB experience on many power stations
both fossil fired and nuclear, is that this procedure saves
considerable time later in the construction programme by
detecting many errors at this early stage. The availability
of a live control desk on the power station with indications,
switches and lamps that respond in a way close to when
connected to real plant offers a valuable opportunity to
familiarise operators and commissioning staff with the system
behaviour prior to real plant commissioning.

Earlier simulators used for closed loop testing were of
the analogue type using operational amplifiers. Later, hybrid
simulators with large patchboards became available and were
used extensively. Recently purely digital systems in the
form of complete proprietary devices have become available.
As an alternative applicable to direct digital control systems,
the plant model can be produced using a high level language
such as CUTLASS [31]. The plant model can be loaded into the
computer and tested "back to back" with the control system
program.

The disadvantage of this purely "software" method is that it does not test all the control system hardware and software. In order to test these, it is necessary to involve the input/ output hardware and software with hardware connections to the desk modules. The plant model is then run in a separate simulator, as shown in Figure 10.

Models used for control loop testing

It must be emphasised that high accuracy of modelling is not required for this testing. Though it may be adequate for the first setting up of control terms, it is not intended for fine tuning and optimisation of the control system. Typically, for the latter, the constants finally set are based on comprehensive modelling of the type discussed in Section 7.1.

XIV. COMPUTER HARDWARE NECESSARY

In order to run large models in real time, large computing power is necessary and a variety of approaches have been adopted, the main ones being as follows:

* Large general purpose mainframes, for example, the IBM 3801 used by CEGB to run its large simulations for design purposes and for the training simulators described in Section VI [5,8].

* Large, general-purpose 32 bit minicomputers.

* Special purpose parallel processors such as the AD10 [32].

* Parallel processors with the plant models being divided and run in separate microprocessors as in the Marconi Graduate System used in the SSEB simulators [6,7].

* Distribution of the computing task into a large number of small processors, as proposed by SINGH and DANIELS [33].

For relatively small models the special purpose parallel approach has been shown to give operation at high speeds and provide real time simulation. Such schemes require specialised mathematical techniques that take the original models and make them operate in the distributed systems. While these are clearly highly efficient, care must be taken to ensure that they are totally valid and that fidelity of modelling is

not lost. Having once validated the modelling process, any
subsequent changes must be very carefully checked to preserve
the quality assurance (QA) route.

By using large general purpose mainframe processors and
high level languages, such as PMSP, the complete plant model
can be constructed from 'macros' [8]. These can be arranged
to have carefully established interfaces so that modifications
and extensions can be closely controlled and the QA preserved.

It would appear that future developments are needed to
make use of the high speed of the specialised architectures
while enabling high level languages to be used with
structures that facilitate development and extension while
preserving the QA. This is particularly important for
established users who have a very large investment in modelling software currently running on large mainframes.

An alternative would be for mainframes with very large
computing power to become available at acceptable cost. These
would probably employ all the relevant techniques including
the parallelism and pipeline operation now exploited in the
special purpose machines. In the medium term, this appears
to be an attractive approach for large models, but technolog-
ical advance is very rapid and other approaches may be more
attractive even in the short and medium term [34].

XV. CONCLUSIONS

A review of the uses of physical and mathematical
modelling techniques in nuclear power station design, engineering and operation indicates that while these have their
own specific purposes they also exercise disciplines which
improve the general systems engineering. An example is the

establishment of a comprehensive computerised database at an early stage in the project, with user-friendly multi-access features.

Considerable advances in the design of the man-machine interface have been made by the use of static full scale mock-ups with or without enhancement with program driven dynamic displays. However, such techniques have limitations. These limitations do not apply if a plant analyser embodying an interactive model is used for this ergonomic development. This technique will permit the development to proceed quickly to a higher level at which there are known severe technical problems. These include alarm displays and their use in an integrated information system.

Plant analysers will have an important role to play in plant operation, specifically in relation to the Technical Support Centre and as a sophisticated operator aid.

ACKNOWLEDGEMENTS

Acknowledgements are due to CEGB, Generation Development and Construction Division for permission to publish this Review. The author also wishes to thank his colleagues for assistance in its preparation, and specifically Mr J.N. Dodd and Mr J.E. Simpson for contributing to Sections VII and XII respectively and to Mr M.J. Whitmarsh-Everiss for Table 1. Some of the material on nuclear plant analysers formed part of a feasibility study, commissioned by the Directorate of Energy of the European Economic Community, and made jointly with Professor D. Zanobetti of the University of Bologna, Italy.

REFERENCES

[1]   STUART, I.F. "BWR Nuclear Plant Maintenance Simulation"
      Inst.Nucl.Eng Conf. on Simulation for Nuclear Reactor
      Technology Paper 3.5. Cambridge 9-11 April, 1984.

[2]   JERVIS, M.W. "Control and Instrumentation of Large
      nuclear power stations" IEE Proceedings Vol. 131 Pt.A
      September 1984 pp 481-515.

[3]   JENKINSON J. "Operator information displays for normal
      operation and fault management of an AGR" IAEA Seminar
      on Diagnosis and Response to abnormal accuracy at nuclear
      power plants. Dresden, June, 1984. Paper IAEA - SR -
      105 - 29.

[4]   CACCIABUE P.C., TOZZI A. "Development of a fast running
      accident analysis computer program for use in a
      simulator". Inst. Nuc. Eng. Conference on Simulation
      for Nuclear Reactor Technology Paper 4.5. Cambridge
      9-11 April 1984.

[5]   BUDD G.C. "The development of full-scope AGR training
      simulators within the CEGB". [This volume]

[6]   McWHIRTER A.F. "The design of mathematical models of the
      nuclear boilers and conventional plant in the Hunterston
      'B' AGR operator training simulator". Inst. Nuc.Eng.
      Conference on Simulation for Nuclear Reactor Technology
      Paper 1.5 Cambridge 9-11 April, 1984.

[7]   HACKING D & HAMILTON J. "Commissioning methods applied
      to the Hunterston 'B' AGR operator training simulator".
      Inst. Nucl. Eng. Conference on Simulation for Nuclear
      Reactor Technology Paper 3.6. Cambridge 9-11 April, 1984.

[8]   CHAMBERS T.H.E. & WHITMARSH-EVERISS M.J. "A methodology
      for the design of plant analysers". [This volume]

[9]   HEYER, D.W. and FRANZON, A. "Conceptual Simulation for
      Interactive Training" Nuclear Eng. Int. June, 1984
      pp 36-42.

[10] ALLAN, R.N., DE OLIVERIA, M.F., KOZLOWSKI, A. AND
     WILLIAMS, G.R. "Evaluating the reliability of electrical
     auxiliary systems in multi-unit generating stations".
     Proc. IEE, 1980, 127 Pt. C., pp 65-71.

[11] ALLAN, R.N., AVOURIS, N.M., KOZLOWSKI, A and WILLIAMS, G.T.
     "Common Mode Failure Analysis in the Reliability
     Evaluation of Electrical Auxiliary Systems". Proc. Third
     International Conference on Reliability of Power Supply
     Systems". IEE September, 1983.

[12] LYNCH, C.A., SMITH, A.A. and EFTHYMIADES, A.E. "Use of
     Interactive Network Graphics Based Power Systems
     Analysis in Distribution Network Operation". IEE Conf.
     on Power System Monitoring and Control, June, 1980
     pp 175-180.

[13] ANCARANI, A and ZANOBETTI, D "Nuclear Plant Analysers :
     their approach to analysis and design". Inst. Nuc. Eng.
     Conference on Simulation for Nuclear Reactor Technology.
     Paper 1.5, Cambridge 9-11 April, 1984.

[14] KAPLAN, G., "Nuclear Power Plant Malfunction Analysis".
     IEEE Spectrum, June 1983. pp 53-58.

[15] ALLEY, A.D., DANIEL, R., KRUGER, G.B., MEYER, R., PUGH, L.
     and QUINN, J.E. "Application of Real Time Interactive
     Engineering Simulation to Advanced Reactor Control
     Design". Inst. Nuc. Eng. Conference on Simulation for
     Nuclear Reactor Technology Paper 1.4. Cambridge 9-11
     April, 1984.

[16] WULFF W., CHENG H.S., LEKACH S.V. and MALLEN A.N.
     "High-speed BWR power plant simulations on the special
     purpose peripheral processor AD10. Inst. Nucl.Eng.
     Conference on Simulation for Nuclear Reactor Technology.
     Paper 1.2. Cambridge 9-11 April, 1984.

[17] KARPPINEN, J., STOKKE E., RINTTILA, E. "A New Dedicated
     Simulator for Man-Machine Systems Research". IAEA/NPPCI
     Specialists Meeting on 'Nuclear Power Plant Training
     Simulators', Otaniemi, Finland. 12-14 September, 1983.

[18] United States Nuclear Regulatory Commission. "Requirement for emergency response capability." Document SECY-82-11, 1982. and NUREG 0696.

[19] SINGH M.G., HASSAN, M.F., CHEN YL.LI, SD. DAN, Q.R. "New Approach to Failure Detection in Large Scale Systems". IEE Proc. Volume 30. Pt. D. No. 5 September, 1983. pp 243-249.

[20] TYLEE, J.L. "On-Line Failure Detection in Nuclear Power Plant Instrumentation". IEEE Transactions on Automatic Control. March, 1983. Vol. AC-28. No. 3 pp 406-415.

[21] HERBERT M.R. "A Review of On-Line Diagnostic Aids for Nuclear Power Plant Operators" Nucl. Energy 1984 23 pp 259-264.

[22] SKINNER, M.S.A. "The role of on-line modelling in detectir and controlling boiler fouling" VGB Conference on 'Slagging, Fouling and Corrosion', Essen February, 1984.

[23] IPSA - Interactive Power System Analysis. A brief description of the facilities and method of operation, published by the UMIST Director of Industrial Liaison, Research and Consultancy Services, P.O. Box 88, Manchester, England.

[24] BRAMELLER, A., JOHN M.N., SCOTT M.R. "Practical Diakoptics for Electrical Networks". Chapman and Hall, 1969.

[25] HORNE, B.E. "The Essential Systems Integrity Monitor". Inst. Nuc. Eng. Conference on Simulation for Nuclear Reactor Technology. Paper 3.4, Cambridge 9-11 April, 198<

[26] BIRCHALL, J.H.L. and CHRIMES, D.A. and EVANS, D. "An emergency reactor cooling simulator at Wylfa Power Station". Inst. Nuc. Eng. Conference on Simulation for Nuclear Reactor Technology Paper 2.2. Cambridge 9-11 April, 1984.

[27] GREEN A.E. and BOURNE A.J. "Reliability Technology" John Wiley & Sons Ltd., London, 1972.

[28] HOLLNAGEL, E. "A Conceptual Framework for the System Description and Analysis of Man-Machine System Integration". I.Chem. E. Symposium No. 50. Ergonomics Problems in Process Operations. EFCE Pub. Series No. 38 pp 81-93.

[29] EMBREY D.E. "Application of Human Reliability Assessment Techniques to Process Plant Design". EFCE Pub. Series No. 38. pp 65-75.

[30] RASMUSSEN J., 1980 "The Human as a System Component". In Smith, H.T. Green, T.R.G. (eds) Human Interaction Computers, (London:Academic Press).

[31] ANNESS D.L. and ROPER P.M. "CUTLASS takes control of more power plant" Electrical Review 209. 29-30, 1981.

[32] WRIGHT, P. "Digital computers in time critical simulation" Data Processing 25. 9 Nov. 1983, pp 30-33.

[33] SINGH, M.G. and DANIELS, B.K. "Rapid Simulation using Low Cost Parallel Computation" Atom 333 July, 1984 pp 11-13.

[34] WILSON, K.G. "Science, Industry and the New Japanese Challenge" IEEE V.72. No. 1 Jan, 1984. pp 6-18.

[35] MYERSCOUGH, P.B. "Role of simulation in the training of nuclear power station operating engineers in the CEGB". Inst. Nuc. Eng. Conference on Simulation for Nuclear Reactor Technology, Paper 2.6. Cambridge 9-11 April, 1984. Also ATOM 334 August 1984 pp 8-10.

[36] HORNE, B.E. and PULLEN, R.A. "The Essential Systems Status Monitor" ANS/ENS Topical Meeting on probabilistic methods and applications" San Francisco, March 1, 1985.

[37] CACCIABUE, P.C., AMENDOLA, A and MANCINI, G. "Accident Simulator Development for Probabilistic Safety Analysis" International ANS/ENS Topical Meeting on Probabilistic Safety Methods and Applications, San Francisco, 24-28 February, 1985.

PSYCHOLOGICAL ASPECTS OF SIMULATOR DESIGN AND USE

R.B. Stammers

Aston University
Birmingham, U.K.

SUMMARY

This paper reviews a number of psychological issues in the simulator area. Three main uses of simulators, from a human factors perspective, are outlined. These are the design of plant and control rooms, performance measurement and training. The paper emphasises the training role of simulators. The role of realism in training devices is discussed as is their use in training for diagnostic skills. The problems of the evaluation of training in a high technology area are outlined and a number of alternative approaches to training are discussed. These alternative approaches include part-task training and computer-assisted learning. A number of other roles for the full scope simulator are also discussed, e.g. such things as the study of advanced skill development and confidence building, the problems of stress and fatigue, the retention of skill and the use of simulators in problem-solving. A clear role for complex simulators is foreseen for the future; at the same time, it is suggested that various alternative forms of training, typically involving the increasing use of computers, are also likely to emerge.

I. INTRODUCTION

The use of simulators and other simulation activities has a history as long as nuclear science and technology. If a broad view of the simulation process is taken so as to include

any desk top activities that attempt to 'represent' a real
process, then it could almost be said that the simulation process
predates any formal activity in science and technology. The
use of some mental representation of a physical system, it
is suggested, is a basic human cognitive activity that must
precede any engineering or design event. A narrower view
of simulation is usually taken, however, with the physical
representation occurring usually within a computer program
and/or working model of a process. If the aim is to represent
the 'knobs and dials' of the real system, then the tendency
is to use the term 'simulator' to refer to a device, usually
representing a control room and/or other associated equipment,
which is used for training purposes. Simulation can retain
a broad meaning referring to such activities as computer
modelling in a science and engineering sense, and in a
psychological sense, referring to any form of representation
of a process with which a human can interact.

Psychological issues in the simulator area come to the
fore because of the essential aspect of simulator use, the
human-system interaction. In this context the focus is
typically upon the operator or control room engineer, who
interacts directly with the system. It is however important
to remember that the design and construction process is also
a human activity and errors can arise in this context as
well as in the control activity. In addition to this, the
whole area of maintenance, corrective and preventive, should
also be considered an important area of human activity.
There has been a continuing but fairly low level of human
factors at work within the nuclear industry for a number of
years, but it is very clear that events in 1979 led to a
large growth in research and development work. This has been
particularly true in the USA but notable developments have
also occurred in Europe and other parts of the world. The
demonstration of the potential for human fallibility at TMI-2
awakened the industry to the need for more human factors
research. Luckily the field was at a stage of maturity that
has been able to respond to the demand and has been able to
make progress on a number of fronts, e.g. improving the
design of control rooms and in the provision of training.
Legislation, political pressures, and a concern for improved
systems has led to a large increase in the number of simulators
provided for training and other purposes. It is worthwhile
therefore to review the more psychological issues involved
at a time when developments are proceeding on the hardware
and software fronts.

II. SIMULATOR USES AND HUMAN FACTORS

Three areas can be considered under the general heading of an interaction between simulators and human factors. The first one is concerned with the design of control rooms, etc. from an ergonomics point of view. In this context, simulators can be used in the development of human system interfaces. This may concern such basic issues as line of sight for instruments, grouping of instruments, design of controls for ease of reach, etc. Also involved are the more complex design decisions that are having to be made on the provision of computer terminals as replacements or additions to conventional instrumentation. In this context, simulations may range from reduced scale mock-ups to full-scale desk mock-ups, from simple computer representations of tasks to realistic control-display interfaces. The goal of this activity is to produce an effective interface for the user. The final stage of such a design process would be the evaluation of performance within some simulation of the real task.

This takes us on to a second major use of simulators which is in the collection of data on human performance. A major concern within the human factors of complex systems is the notion of human reliability (Meister, 1984). The focus of this work is on providing estimates of the likelihood of human error in various decision and control actions. The field is a developing one, but progress has been made over the last two decades. More recently, the idea has been put forward that operator performance in simulators could be used as one way of collecting data on human reliability. Recent studies in this area (e.g. Beare and Dorris, 1983, 1984) have demonstrated the feasibility of the technique and have shown that, in the main, results obtained from simulation exercises have approximated those generated from other forms of exercise. In addition, these studies have enabled such variables as length of experience and the role of technical advisers to be studied in a more objective way, with the capability of collecting usable human factors data. This role of a simulator makes particular demands upon the realism of the simulator activity and whilst some strides can be made towards an assumption of reality, it of course will always remain, in the minds of the operators, a simulation exercise. A related role for the simulator is that of performance assessment and of testing. The need in some countries for an annual licensing of operators presents

problems in terms of how competence can be assessed. The
realistic simulator is suggested as one way in which such
assessment can be done. This again places particular demands
on realism of simulators and their correspondence to the
plant on which the operator is expected to function. These
demands on realism for performance assessment purposes are
potentially very different from those required for training
purposes. This topic will be discussed at greater length
below but it is important to bear in mind that realism for
one purpose may not be needed for another role.

This brings us to the final use of simulators, that of
training. This topic will be dealt with at greater length
in this paper. Questions are raised on of what role the
simulator has to play and what are current states of developments of alternatives and adjuncts to the provision of full-
scale simulators for training purposes. The use of simulators
in the nuclear industry for training purposes has a long
history (Stammers, 1985) but, as mentioned above, the topic
has more recently received greater attention. There has been
a subtle shift of emphasis away from the provision of a
limited number of simulators at a central training facility
towards additional provision of simulators on site. The
role of the on-site simulator is not just to provide an
advanced training facility but also to be used as an
exploratory device for testing out hypotheses about real
plant states during incidents.

III. SIMULATORS AS LEARNING ENVIRONMENTS

The principle aim of a simulator should be to provide
an efficient and effective learning environment. The
principles and procedures for the design of instructional
systems have been well established (Stammers, 1984). Whilst
there is some debate over details, the broad nature of training
technology has been clearly laid down. Thus for training for
process control skills there is little debate over the need
for some form of plant simulation. On the other hand,
questions of detail concerning how realistic the plant has to
be simulated remain controversial. Similarly, it is recognised that a healthy mixture of theory and practice is required
to successfully operate complex plant. However, the most
appropriate mix of theory and practice has not been established.
Certainly, the effectiveness of a large amount of basic theory
has been seriously questioned (Duncan, 1981). There is no
guarantee that abstract theory, however well taught and however

effectively learnt, will transfer to effective performance
on the plant. Similarly, learning on-the-job or by exposure
and experience is likely to be an inefficient method. The
author has been led to query whether very often the drive for
realistic simulators arises from the idea that if people can
learn effectively on the simulator, no matter whether we
understand what is going on, this avoids the need for a
detailed task analysis and training program design. It is
certainly possible to learn the complexities of the task in
a simulator in an inefficient way; it is hoped that an
appropriately structured training programme will enable this
to occur even more efficiently.

   The detailed theoretical background to the determination
of training content has been discussed elsewhere (Stammers,
1985). Suffice it to say here that knowledge in the form of
knowing what is going in the plant, and knowledge (or skill)
of how to operate the plant is what should be the basis of
any training programme. In operating complex plant it is
likely that procedural skills in the terms of correctly
identifying displays and controls and correctly operating
them in the right sequence, will form the basis of a great
deal of the required performance. It is also likely that
such procedural performance will be supported by job aids,
procedural guides and the like. More importantly, therefore,
remain the diagnostic skills of the human operator as the
ultimate back-up system. These diagnostic skills have been
isolated by a number of writers as of key importance in the
control of complex plant. For example, in the context of
chemical processes, Marshall et al.,(1981) see diagnostic
skills as highly relevant in continuous control processes.
These skills need to be of a general and specific nature
given the complex variety of process plant that the operator
is likely to be confronted with. Similarly, Johnson and
Maddox (1983) carried out a survey in the US nuclear industry
which pinpointed the importance of diagnostic training for
power station operators. In a wide ranging survey, this
topic emerged as the key training issue. In relation to this,
Thames (1985) suggested that whilst a proliferation of simulators had followed the TMI events, very little attention
had been paid to the specific provision of training aids for
diagnostic skills. This would fit with the idea put forward
above, which might be termed the 'principle of inclusion'.
The idea being that if the complex simulator exactly replicates
all of the plant characteristics then all possible plant
states can be represented upon it. Whilst this is possible,

it appears to be an inefficient way to go about tackling a
particular issue. A central difficulty for work in the
continuous process area is that of evaluation of training.
The difficulty arises from the fact that assessments of
performance in the real situation are difficult or near
impossible. Traditional training models are based on a
transfer from training to some production or some other
measurable activity such as vehicle or weapon system control.
Different forms of training can be evaluated by means of
'transfer of training' exercises. Transfer may be measured
on the basis of initial performance or in some cases may be
measured on the basis of the amount of learning on the job
required to reach an acceptable standard. In process
control and other areas of high technology (e.g. aviation),
it is very rarely possible to directly measure performance.
It is unacceptable to allow errors to be made within such
systems. Therefore, trainees will be closely monitored and
supervised during their initial performance of tasks. This
does of course offer some scope for performance measurement,
but is again not very effective. Another problem is that
systems are, hopefully, inherently reliable. Therefore the
need for human skill intervention is typically rare. Thus,
training received at one point in time may not be utilised
until some later date or indeed, may not ever be called upon
during the working life of one individual. There are problems
therefore for the industry in terms of performance assessment.
This suggests that the full scope simulator has a role to play,
not just in training, but in the assessment of that training
by measuring performance in realistic environments. Indeed,
it might be suggested, in the light of what is to follow, that
this emerges as the key role for simulators in the future.
Various alternative forms of training technology may be brought
to bear on the 'learning problems' of the individual, rather
than upon their 'performance of skills' activities. The aim
is not to denigrate advanced simulators but rather to
identify for them their most effective role in advanced
training and performance testing, and to look for alternative
methods for initial training and development of skills.

IV. ALTERNATIVE APPROACHES TO TRAINING

Alternative approaches to training for complex nuclear
plant control that suggests themselves are, in the main,
techniques that abstract from the complexity of the real
situation essential elements for representation on training
devices. Sometimes these abstractions take the form of a

simplification of the situation. In other cases, they involve
the isolation of a particular part of the task or provide the
opportunity for the practice of a particular skill from within
the complex of those required for plant operation.

Looking firstly at the question of abstraction, one
particular form of training device that has proved popular is
the concept or principle trainer. Various of these have been
developed for the nuclear industry (e.g. Chadwick and Walker,
1985). Their use is typically in initial training. The role
is to introduce trainees to the nature of the processes they
will be controlling, the way in which various components
interact, the nature of the chemical and physical processes
involved, and provide illustrations of the consequences of
particular actions or inaction. The aim is a laudable one,
that of linking new knowledge, in a generalised form, to
previously existing knowledge. This knowledge should also
form the basis for further elaboration by the more detailed
training processes that are to follow. Such training devices
also have a role to play for ancilliary personnel who may
require an understanding of nuclear power processing without
requiring the detail provided by full scope simulation. Such
facilities may be provided by a general purpose principle
training or may be provided as part of a computer-based
training package, the plant being represented on a computer
screen. It is also important to note that such general purpose
training facilities will have value across a range of plants
if the principles represented are general enough. This is
in contrast to full scope simulators which very often can only
represent one plant in any detail. The feasibility therefore
for portability either in terms of the device or a particular
software package, is attractive.

The provision of reduced fidelity simulators is now
emerging in a number of centres in the context of computer-
based training (CBT). As well as training being available
on well established computer-assisted learning systems (e.g.
PLATO), a number of other projects are pursuing this line.
A good example is the work of Johnson et al., (1984). Their
work provides an illustration of diagnostic skill training in
a demonstration system. The system provides, on a visual
display, a representation of the plant, tasks the individuals
with diagnostic activities to be carried out and then provides
very detailed feedback to the learners on their performance.
The aim of the work is to yield a system that is very easily
learned and related to the real system, but which places an

emphasis on trouble-shooting skills. Another system under
development takes basic ideas of simulation and ties them to
emerging ideas of intelligent tutoring systems. This is the
project called 'STEAMER', a CBT system being developed for the
US Navy (Williams et al., 1981). This research attempts to
represent a steam propulsion system on a desk top computer.
The representation of the system is to be linked to an intelligent tutoring device which will guide the learner through
the intricacies of the plant. These researchers, like others
in the area, see an importance in the learner grasping a
conceptual understanding of the processes involved rather than
in the rote learning of specific sets of drills for specific
circumstances. In order to do this they envisage a highly
flexible interaction between the learner and the computer
terminal, such that the learner is able to explore the complexities of the plant with appropriate guidance and feedback
from the instructional system. Whilst a mathematical model
of the process forms for the basis of this approach, the
various instructional strategies developed will have general
application across a range of different types of plant.
Similar ideas have already been developed in the context of
electronics trouble shooting (Brown et al., 1975).

It is not necessary, however, to always follow a 'high
technology' route towards tackling such training problems.
An inexpensive system is described by Marshall et al. (1981).
This makes use of simpler technology and places the key
tutoring role in human hands, those of the instructor. In
their system, photographs of simulated plant conditions are
back projected on to a screen to yield life-size representations
of the panel. Trainees work through various exercises to
learn generalisable diagnostic rules. The authors suggest
that the learning of such rules has much to commend it when
a person is confronted with a plant of any degree of complexity.
Whilst specific rules for specific circumstances will emerge,
their suggestion, backed up to some extent by empirical
evidence, is that generalisable rules will 'transfer' to novel
situations. This offers some promise for training techniques
that have some degree of generalisability to complex situations.
Their system, whilst labour intensive in terms of production
and running, requires fairly simple technology. Computer-based
training systems described above, whilst engaging in their
flexibility and contemporary appeal, will have large software
development implications which need to be taken into account.
It is likely however, that general purpose training systems
will emerge which will have the ability of linking instructional

strategy programs to simulations of particular plants. In our
own laboratory, we are developing such a training approach
which we have termed 'adjunct training'. The aim is to build
a link between computer-based training packages with general
teaching strategies and complex simulations which in themselves
constitute highly complex computer programs. The training
package drives the simulation through various states with
data passing between the two programs. The trainee is able
to work at the instructional workstation, where basic
instruction is given and introductory rules and principles
supplied. The trainee then transfers to the task simulation
interface and carries out specific tasks. Data on the adequacy
of responses, etc. is passed back to the instructional control
program which initiates appropriate further training exercises.

The final topic that can be considered in this area is
the development of part task trainers. It could be suggested
that any trainer that simplifies the real task could be
considered a part task trainer, and therefore into this
category would fit the principle trainers described above.
However, part task trainers are usually taken to mean trainers
where a high degree of reality is presented but for only part
of the system, therefore, only parts of the task are represented.
Thus components of the plant might be realistically represented
in single training devices (e.g. Birchall, 1985). Such
individual training devices will enable parts of the task to
be practiced in isolation. This may make sense if those
tasks are only ever performed in such a form. If, however,
such tasks at some stage need to be integrated with other
skills, then there is a need to handle the transfer from the
part task to the whole task situation. This may be feasible
on transfer to the real situation where the trainees can be given
supervision and monitoring such that they are able to integrate
their part task skills without the need for the full scope
simulator. Whilst in some areas this will be acceptable, in
others the degree of coordination required might in turn
suggest the need for a full scope as well as part task trainers.

V. REMAINING TASKS FOR THE FULL-SCOPE SIMULATOR

The above section outlines various alternative approaches
to the development of knowledge and skill for nuclear plant
operators. Many of these training activities are possible
within the full scope simulator, but it is a question open to
research as to whether these tasks can be carried out more
effectively in other training situations or just as importantly

more cost effectively in reduced fidelity situations. It is suggested that these research questions are amenable to investigation, given appropriate resourcing. The issue is that such research is only possible within the context of existing full scope simulators. It is also suggested that we are unlikely to see the demise of full scope simulators in this industry in the near future, even if alternative approaches for basic training were shown to be effective. A number of tasks remain for the full scope simulator. Already described in a previous section is their role in the testing and/or licensing of operators. Until alternative methods are found or the validity of a reduced scope simulator to produce effective performance measurement is established, the devices are likely to remain in this role. In addition, they are likely to remain as a result of political or legislative pressure as a way of ensuring and boosting operator competence. Also until alternatives are found they will be retained because of their high face validity as effective training environments.

Another relatively unexplored area also requires full scope simulators as its basis. This concerns such topics as skill development and confidence building. It is suggested that the full scope simulator offers more opportunity for the development of high levels of competence by operators and in turn, a degree of self confidence in the skill possessed. Strong pressures exist within the aircraft industry for a high degree of realism in flight simulators. One suggestion as to why this is so is that in such simulators, pilots are able to assess their own confidence and are able to see their skill tested in a number of situations and see it refined and extended to cover a wide range of opportunities. It is likely that the same situation exists within the nuclear industry. The higher the perceived degree of reality of the simulator, the higher level of confidence operators are able to attach to their perceived level of skill.

Related to the above topic is the question of stress and fatigue. Another human factors issue not mentioned above concerns the effect of stress and fatigue on performance. Given that emergencies may in themselves give rise to stress on the part of the operator, deriving either from the fact that their competence in the situation will be assessed or that the situation itself may be personally or environmentally damaging, then it is likely that the performance of the individual is likely to be affected by stress. Add to this, that such incidents may arise at any time within the twenty

four hour shift cycle, then it is also possible that problems
of fatigue, etc. may be present. The effect of stress and
fatigue on human performance is a complex one, and is a topic
which is still the subject of intensive research (e.g. Hockey,
1983). It is fairly clear that under a number of situations
performance in the presence of environmental stressors may
degrade performance. The issue is complicated however as,
under certain circumstances, the presence of a stress may
indeed enhance performance. Performance in the presence of
two or more stressors may have an enhancing, a negative or a
neutral effect on performance. The majority of studies concern
laboratory task performance, carried out in the presence of such
stressors as loud noise, loss of sleep, ingestion of alcohol,
etc. Whilst these have some relevance to performance in the
control room context, it can be pointed out that stress here
arises from the task situation itself. It is likely that
performance under conditions of fear or within hazardous
environments is likely to have a negative effect on performance.
For example, studies reported by Izikowski and Baddeley (1983)
on the performance of divers, suggest that performance can be
degraded significantly when divers carry out their tasks in
bad sea conditions and at depth. Performance under simulated
conditions in pressure chambers is not so affected. The
degradation of performance therefore is likely to be due to
the performance within the hazardous environment itself
rather than as a result of the physical stressors (e.g.
pressure) impinging on the individual. These authors also
report on studies carried out in a military context under
either true or simulated combat conditions. In such situations,
although performance measurement may be difficult, degradation
in performance is likely to be found (Berkun, et al., 1962).
The stress condition therefore is a complex one and can be
approached in a number of ways. The full scale simulation
approach is typically one that says, let us make the situation
as much like the real life as possible in order to engender a
feeling of stress, etc. as close as possible to the real life
task. Whilst this is a laudable aim, it is difficult to
evaluate in any situation other than one where the trainees
may be deceived into believing that they are carrying out the
task on the real equipment when in fact they are working
within a simulator. Otherwise it is always subject to the
criticism that it is a simulation and the operator knows
this. On the other hand, it is fair to say that people can
demonstrate physical symptoms of stress when being put under
pressure in simulated task environments. Whether this arises
from the same form of stress that they will encounter in the

real situation or whether it arises from the social pressure
of being assessed in the presence of trainers, supervisors
and/or other trainees, is an open question.

An alternative approach to building resistance to stress
is that of over training. Here extensive practice is given
on the tasks to be performed. This may be done in a full
scale simulator or it may be more cost effectively done in a
reduced scope simulator. Under such conditions, intensive
practice with a range of fault conditions can be given to
trainees. Duncan and Shepherd (1975) describe such a system.

Such training approaches also have relevance to another
area of concern within the industry. This concerns the
retention of training. It has already been mentioned above
that due to the inherent reliability of plant, many of the
skills learned during simulator practice will be little
practised in real life. Given that practice is one of the
most potent variables leading to the retention of skill it is
not surprising that there is a concern with the extent to
which retraining should be given and how often. Low fidelity
training devices offer promise in this area for effective
rehearsal of skills. The retention of skill area has been
reviewed on a number of occasions (e.g. Stammers, 1981).
Evidence exists that skills not practiced will be lost.
However, relearning can be rapid and can often occur with
reduced fidelity simulation. Whilst much of the data is
somewhat dated now, recent research (e.g. Hagman and Rose,
1983) has tended to support the earlier findings of skill loss.
Whilst few of the tasks concerned have much resemblance to
the diagnostic skills of process control, there is clearly
the need for this topic to be taken more seriously. One
suggestion is that an investment in overlearning in generalisable skills will pay off as such skills tend to be
inherently better retained over time and can be shown to
transfer to new situations (Ducan and Shepherd, 1975). The
suggestion is therefore that overlearning of generalisable
diagnostic skills, albeit in low fidelity training situations,
is likely to pay off in the long run. Another suggestion is
that such training facilities may be provided in control
room situations. During periods of low activity, operators
may be able to 'test their skill' against a computer in a
game-type situation. Given that such situations would not be
overtly or covertly monitored by management, they may provide
a situation sufficiently motivating for operators to feel
comfortable using them and at the same time, value their
role in enhancing or refreshing their skills.

A final important role for full scope realistic simulators that has been identified is their use in plant problem solving activities. It is envisaged that in the time available during the development of a critical incident, a plant simulator could be put into similar state and various forms of remedial action tried out on it, presumably in fast time. The idea here is that the simulator, either on site, or in direct communication with the control room, will enable a number of courses of action to be attempted. This is particularly important given the nature of human problem solving e.g. its conservatism, the likelihood of a consensus ideas emerging and the possibility of making the situation worse through incorrect diagnosis. In this sense, it might be helpful if computer assistance in the form of decision making expertise could also be provided such that a wide range of alternatives could be assessed and matched against the real situation.

VI. CONCLUSIONS

A number of psychological issues emerge in the design and use of simulators. Although this paper has concentrated on the training role of such devices, it should also have made clear their important role in performance measurement, design and various other activities. It is likely that the full scope simulator will remain an important aspect of this high technology area as it has in aviation. It should also have been made clear that the full scope simulator has a particular role to play and may have limitations in the extent to which it can be a comprehensive training device. It needs to be supported by various other forms of training activity to make the best use of equipment, such that all its complexity is effectively utilised in advanced training, rehearsal training or testing activities. It is important therefore to envisage a training system composed of a number of training activities leading to greater and greater elaboration of skill. The end product of this is the operator who is trained to have a range of diagnostic and procedural skills that can be applied to the plant, that are flexible as the plant develops and enable an easy transfer between different plant configurations. The use of low fidelity training devices should not be denigrated if their particular role can be identified and shown to be of importance and validity in the development of high levels of skill. Computer technology is likely to have an increasing part to play in the provision of training. This might be considered under the heading of low fidelity training devices but this would

only be in terms of the physical representation of the plant.
The complexity of such devices would lie within their flexible
and adaptive training strategies. Research in this area is
likely to yield general purpose training philosophies that
can be linked to various forms of plant simulation.

It is felt that existing psychological expertise is
only now beginning to be used effectively in the industry,
and much more use could be made of human factors research.
On the other hand, a number of research questions remain
which can only be answered by the close collaboration between
nuclear scientists, engineers, plant management, operators
and human factors specialists.

REFERENCES

BEARE, A.N. & DORRIS, R.E. (1983). 'A simulation-based
study of human errors in nuclear power plant control
room tasks'. In: Proceedings of the Human Factors
Society, 27th Annual Meeting. (Santa Monica, CA: Human
Factors Society). pp. 170-174.

BEARE, A.N. & DORRIS, R.E. (1984). 'The effects of
supervisor experience and the presence of a shift
technical advisor on the performance of two-man crews
in a nuclear power plant simulation'. In: Proceedings
of the Human Factors Society, 28th Annual Meeting.
(Santa Monica, CA: Human Factors Society). pp. 242-246.

BERKUN, M.M., BIALEK, H.M., KERN, R.P. & YOGI, K. (1962)
'Experimental studies of psychological stress in man'.
Psychological Monographs, 76, No. 15.

BIRCHALL, J.H.L., EVANS, D. & CHRIMES, D.A. (1975). 'An
emergency reactor cooling simulator at Wylfa power station'
In: Walton, D.G. (ed.) Simulation for Nuclear Reactor
Technology (Cambridge: University Press) pp. 117-135.

BROWN, J.S., BURTON, R.R. & BELL, A.G. (1975) SOPHIE.
'A step toward creating a reactive learning environment.
International Journal of Man-Machine Studies, 7, pp.
675-696.

CHADWICK, G & WALKER, S (1985). 'The evaluation and subsequent redesign of a basic principle training simulator'. In: Walton, D.G. (ed.) Simulation for Nuclear Reactor Technology (Cambridge: University Press) pp. 111-116.

DUNCAN, K.D. (1981) 'Training for fault diagnosis in industrial process plant'. In: Rasmussen, J & Rouse, W.B. (eds.) Human Detection and Diagnosis of System Failures (New York: Plenum) pp. 553-573.

DUNCAN, K.D. & SHEPHERD, A. (1975). 'A simulator and training technique for diagnosing plant failures from control panels'. Ergonomics, 18, pp. 627-641.

HAGMAN, J.D. & ROSE, A.M. (1983). 'Retention of military tasks: A review'. Human Factors, 25, pp. 199-213.

HOCKEY, R. (1983, ed.) 'Stress and Fatigue in Human Performance'. (Chichester: Wiley).

IDZIKOWSKI, C & BADDELEY, A.D. (1983). 'Fear and dangerous environments'. In: Hockey, R. (ed.) Stress and Fatigue in Human Performance. (Chichester: Wiley). pp. 123-144.

JOHNSON, W.B. & MADDOX, M.E. (1983). 'Status of diagnostic training in the nuclear utility industry'. In: Proceedings of the Human Factors Society 27th Annual Meeting. (Santa Monica, CA: Human Factors Society) pp. 151-155.

JOHNSON, W.B. & MADDOX, M.E. & KIEL, G.C. (1984) 'Simulation-orientated computer-based instruction for training nuclear power plant personnel'. In: Proceedings of the Human Factors Society 28th Annual Meeting. (Santa Monica, CA: Human Factors Society) pp. 1008-1012.

MARSHALL, E.C., SCANLON, K.E., SHEPHERD, A & DUNCAN, K.D. (1981) 'Panel diagnosis training for major-hazard continuous-process installations'. The Chemical Engineer, No. 365, pp. 66-69.

MEISTER, D (1984). 'Human reliability'. In: Muckler, F.A. (ed.) Human Factors Reviews: 1984. (Santa Monica, CA: Human Factors Society) pp. 13-53.

STAMMERS, R.B. (1981). 'Skill retention and control room operator competency'. (Karlstad, Sweden: Ergonomrad, AB) Rep. No. 19.

STAMMERS, R.B. (1984) 'Current developments in training technology'. In: Proceedings of the International Conference on Occupational Ergonomics, Toronto, Volume 2. (Rexdale, Ontario: Human Factors Association of Canada) pp. 163-170.

STAMMERS, R.B. (1985). 'Instructional psychology and the design of training simulators'. In: Walton, D.G. (ed.) Simulation for Nuclear Reactor Technology. (Cambridge: University Press). pp. 161-176.

THAMES, O.E. (1985). 'The present generation of BWR training simulators - have they kept pace with new trends in training?'. In: Walton, D.G. (ed.) Simulation for Nuclear Reactor Technology. (Cambridge: University Press) pp. 151-158.

WILLIAMS, M., HOLLAN, J & STEVENS A. (1981). 'An overview of STEAMER: An advanced computer-assisted instruction system for propulsion engineering.Behaviour Research Methods and Instrumentation, 13, pp. 35-90.

THE DEVELOPMENT OF FULL-SCOPE AGR

TRAINING SIMULATORS WITHIN THE CEGB

G.C. Budd

Central Electricity Generating Board, U.K.

SUMMARY

This article describes the approach which has been adopted by the Central Electricity Generating Board (CEGB) in developing replica simulators for training plant operations staff from their Advanced Gas-Cooled Reactor (AGR) power stations. A number of important aspects in the development of these simulators are discussed initially, leading on to some specific points related to the AGR Simulator designs themselves. This includes the feature of running the simulator total plant models on general purpose mainframe computers located about 200 km remote from the Training Centre where the simulated unit control desks and panels are actually sited.

I. INTRODUCTION

The CEGB has accumulated considerable experience in developing full-scope digital simulators for power station operational training, both for nuclear and conventional plant. Since 1976, when the first of the CEGB's four AGR Simulators was conceived, the procurement has been a total "in-house" activity, involving a number of departments - viz:

EDUCATION AND TRAINING - the "User".

COMPUTING AND INFORMATION SYSTEMS - the "Software Supplier".

GENERATION DEVELOPMENT AND CONSTRUCTION - suppliers of reference codes and later of overall Project Management and hardware procurement.

133

In presenting the design and development strategy employed for the CEGB's Training Simulators, the following points will be addressed:

(a) User Requirements and detailed Functional Specification.

(b) Choice of plant model complexity to satisfy (a).

(c) Choice of display and tutorial facilities to satisfy (a).

(d) Assessment of likely computer loading, leading to the choice of computer and control desk/panel interface hardware.

(e) Software development strategy including use of standard CEGB packages, and software security considerations.

(f) Long-term support, maintenance and enhancement of the total system (both hardware and software).

(g) Quality Assurance, including validation of plant model software.

(h) AGR Simulator system design, embodying (a)-(g).

The CEGB policy to simulator procurement is based on meeting a training need, so the development process is a naturally iterative operation involving a balance between the ideal training requirements and what is feasible to supply. The scale of the task is illustrated by the typical schematic diagram of major plant items to be simulated, shown in Figure 1.

II. USER REQUIREMENTS

The starting point in any development project of this nature is the end-user's requirement. In the case of the CEGB's AGR Simulators, this is produced by the Education & Training Department, and comprises a comprehensive document covering the following essential items:

(a) The PURPOSE of the simulator (e.g. training only, or including other aspects such as station pre-commissioning trials).

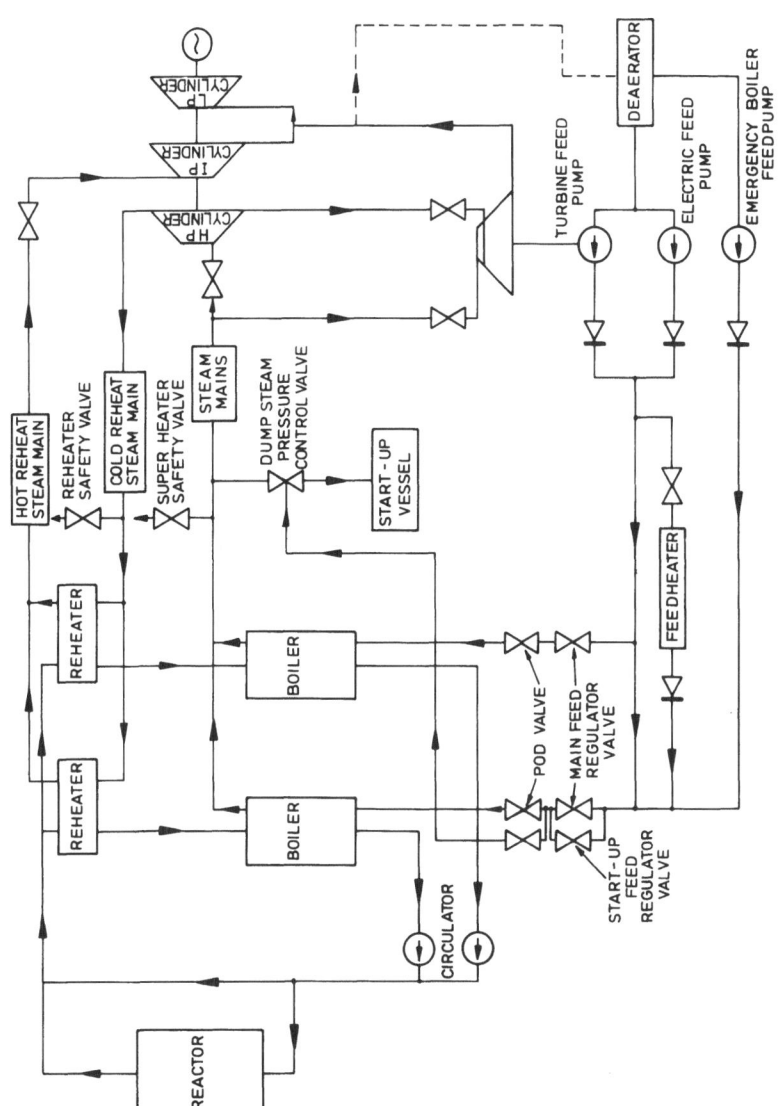

Figure 1

A typical AGR Simulator Schematic Plant Diagram

(b)   The SCOPE of the simulator (e.g. generic or replica).

(c)   The DEPTH OF PLANT to be simulated, with accompanying
      schematic diagrams where possible. This indicates the
      level of plant detail required for training (e.g. all
      reactor auxiliary cooling systems, all dump steam vessels
      and associated controllers, but excluding drains flash
      vessels and/or condensers).

(d)   Typical PLANT OPERATIONS for which training is to be
      provided, which define the range of operation required
      of the simulator.

(e)   FAULT INJECTIONS and OVERRIDES required. These items
      appear simply as a list (or in context) in the User
      Specification, but are expanded in the detailed Functional
      Specification to indicate methods and conditions of
      application and removal.

(f)   Any SPECIAL REQUIREMENTS, such as computer aided assess-
      ment, some of which may eventually be deleted if they
      are not cost-effective or feasible to provide.

(g)   The ACCEPTANCE CRITERION for the simulator. A formal
      acceptance test schedule is arrived at partly by experience
      when simulating a large number of plant operations, but
      the functional testing of desk switches, input/output
      etc., is able to be defined more explicitly. An indication
      of the acceptance criterion does give the supplier a
      "feel" for what the user is looking for overall.

One problem with writing a User Specification for a
replica simulator occurs when that simulator is needed for
training in advance of raising power on the actual station.
In such a case, the plant design is unlikely to be finalised
at an early enough stage to realistically allow the simulator
specification to be frozen.

This was the case with all the CEGB's replica AGR
Simulators, and was one reason why they were developed in-
house, because it enabled the specifications to be retained
in step with the station designs for as long as possible
during development, thereafter upgrading the systems to match
operational and tutorial changes.

Any such changes had to be justified as essential to the
initial training need or able to be incorporated within the

original budget. As an in-house activity, this afforded
considerable flexibility when taking those decisions.

It is a false economy to build a replica simulator to
time and cost if the end-product is so out of step with the
station that its training value is seriously diminished.

III. FUNCTIONAL SPECIFICATION

The form of the detailed Functional Specification for a
CEGB AGR Simulator has evolved to the point where it is now
a comprehensive set of "Design Intent Statements" (DISs)
covering all aspects of the development process. Each DIS
comprises a number of specific sections, and an example of
the structure related to a plant model software system is as
follows:

Section 1 : "Introduction"

1.1 General Requirements - major plant areas covered, and
     outline description of the plant system.

1.2 Sub-Systems Covered - a list of system divisions.

1.3 Assumptions and Limitations - difference between the
     simulation and the physical system.

Section 2 : "Functional Requirements"

2.1 Sub-System No. 1 - area of plant.

     2.1.1 Description of Sub-System - detailed description
            of the plant and its operation.

     2.1.2 Simulator Operational Requirements - limits of
            performance, operational functions to be catered
            for by the simulator (i.e. main simulator
            requirements, acceptable constraints, and
            required range of operation).

     2.1.3 Tutorial Requirements - VDU formats, fault
            scenarios outside normal operational range, and
            other requirements specific to this sub-system.

     2.1.4 Software Implementation - software supplier's
            statement on the manner in which the requirements

will be implemented, giving a broad idea of the
methods to be adopted, with cross-references to
software documentation where appropriate.
No detailed information is required here, unless
it is appropriate (e.g. for the simulation of
novel areas of plant).

2.2 Sub-System No. 2, etc.... (2.2.1-2.2.4).

(Repeated for each sub-system)

Section 3 : "References to Source Information and Schedules"

This section defines the sources of information used in
the preparation of the specification, and also refers
to documents, schedules and data lists required in the
design process.

Only references are included in this section - drawings,
schedules etc., of immediate relevance are contained in
appendices.

Some examples of referenced items would be plant
schematic drawings, control system and interlock schedules,
interface schedules, and extracts from plant operating
instructions.

Section 4 : "Validation"

4.1 Design study data, reference steady states, transients.

4.2 Actual plant data (if available).

4,3 Commissioning/Acceptance test schedules.

Section 5 : "Special Information"

Any other relevant information not specifically covered
in Sections 1-4.

Section 6 : "Appendices"

Plant and simulator schematic diagrams, schedules, etc.
- generally, those items of immediate relevance
referenced in Section 3.

Throughout the course of a simulator development project, the DISs become more detailed as the volume of information increases. In order to formalise the updating procedure, each DIS evolves by way of four well-defined major "versions" as the project proceeds. Each of these versions can itself contain a number of updates, identified by the decimal part (.x) of the version number - viz:

Version 1(.x) - Outline specification only. This contains information relating to plant design and operating procedures, together with tutor requirements for the simulator, at the very early stages of the project. As well as defining some basic information, it also serves as a useful "teaching" document for all those involved in the simulator development.

Version 2(.x) - Specification including input/output (i/o) lists, tutorial requirements, plant data, etc. This version usually goes through many "decimal" updates, simply because detailed information for the final simulator specification is continually being collected during this period.

Version 3(.x) - Specification package complete - plant and control desk drawings, i/o schedule lists with nomenclature, software supplier's response to the design intent, and logic specifications where appropriate (such as for sequences and interlock handling). This includes all information necessary to design the software. Formal project change control starts at this point.

Version 4(.x) - As finally implemented.

It can be seen that the issue of Version 3 of each DIS is a key event in the project timescale. Ideally, the specification would be frozen at this point. In reality, though, further updates always come to light at a later date, so these are considered in relation to their training value as part of the formal project change control procedure. Only if they are deemed essential to training, bearing in mind the cost and their effect on project timescale, are they allowed to go through as a change to the specification. Otherwise, they are documented for consideration as a possible enhancement after the simulator has been formally accepted for initial training.

Any changes to the simulator which come about during long-term support and maintenance (i.e. after formal acceptance) will be noted in "decimal" updates of Version 4 of the appropriate DISs, which then remain as a permanent record of the project.

IV. CHOICE OF PLANT MODEL COMPLEXITY

If an effective and reliable system is to be achieved, then a compromise has to be made between high accuracy and robustness of the numerical solution. It can be argued that it is more important for the simulator to keep running, whilst giving credible results, than to achieve high fidelity within tight constraints. A clear example is when the simulator is driven outside the normal plant operating range due to maloperation, when it must be capable of being driven back into range again without giving obviously erroneous indications.

With this in mind, the attraction of using design codes for Training Simulators is currently tempered by two main factors:

(a)   Their range of valid operation is usually insufficient for the Training Simulator application (e.g. cold start). They are generally intended for specialised studies such as the evaluation and optimisation of control system performance, rather than the wider aspects of total plant operation.

(b)   Their inherent complexity invariably results in eigen-values of large magnitude, and also equations involving implicit algebra. The former preclude the unconditional use of simple fixed-step explicit integration algorithms (e.g. Euler), thereby forcing alternative methods to be used such as implicit or variable-step integration, or local eigenvalue protection [2], all of which involve extra computation. Implicit algebra generally involves iteration, which also uses up valuable CPU time. Consequently, design codes do not always lend themselves to a guaranteed real-time solution, unless the accuracy criteria are significantly relaxed (whence they are no longer design codes).

In general, the spread of plant model detail is different for a Training Simulator than for a design code, because the former will need representations of items such as partial air

volumes, drains for pipe warming, and turbine supervisory
parameters (i.e. shaft eccentricities, bearing vibrations and
shaft-casing differential expansions), whereas the latter will
usually omit these in favour of more detailed reactor or boiler
models for example.

The choice of plant model complexity is driven by a
proper understanding of the training requirement, and a
knowledge of how this can best be achieved by excluding items
of secondary importance from the simulation (if they carry a
penalty in terms of execution speed or solution complexity).
Experience is needed to achieve this aim, and a full technical
exposition of the approach adopted by CEGB modelling staff is
given in [2].

V. CHOICE OF DISPLAY AND TUTORIAL FACILITIES

The CEGB's AGR stations each have a sophisticated Data
Processing System (DPS) for monitoring the plant state and
generating any associated alarms. The DPS is usually a
powerful supermini computer configuration which continually
monitors the plant, and displays alarms and other information
to the operator via one or more VDUs.

The AGR Training Simulators represent these systems
either by "simulation" or"stimulation". In the case of the
early AGR Simulators, the DPS was simulated on a small minicomputer driving a graphical display system. However, the
Heysham II AGR Simulator (currently in development) will
employ replica DPS hardware and software, which will be
stimulated by a minicomputer transmitting the required signals
from the plant model. This option allows future updates of
the station DPS software to also be used on the simulator.

In addition to the DPS, the simulator can provide VDU
displays (formats) of information generated by the plant
model but which would not normally be available on the actual
plant. Some examples are tutor-designed schematic diagrams
(Figure 2), and boiler temperature profiles.

The library of such formats can be quite extensive, but
only a small number can realistically be displayed simultan-
eously. In practice, more than about four or six will impose
an unacceptable extra load on the processor, so this is a
typical number to aim for.

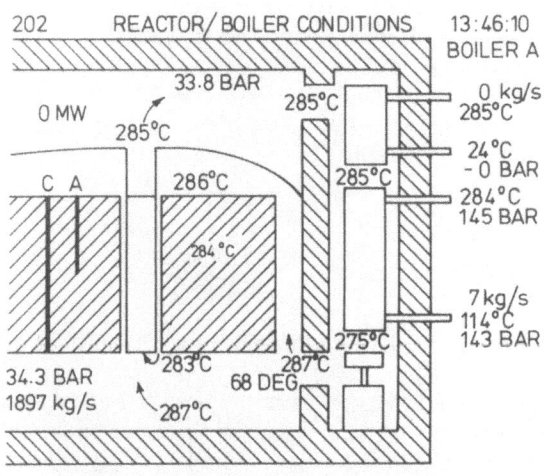

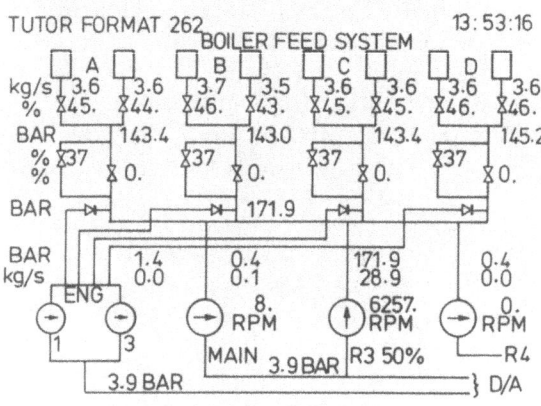

Figure 2

Examples of Tutor Formats

The degree of user-friendliness of the tutor facilities to control the simulator also has an impact on processor load. A balance has to be struck between assuming that the tutor knows nothing about the simulator system structure (and therefore requires very detailed prompts - usually resulting in a significant processor load and storage), and on the other hand assuming a limited knowledge of the system.

More importantly, the sheer volume of information available to the tutor needs very careful organisation and filtering. Therefore, the more sophisticated the tutor facilities become, and the larger the number of factors over which he is given control, then the greater the processing capability is required, and the higher the cost. It is therefore important that the tutor and the supplier both appreciate what is really required, and what is cost-effective to provide.

The tutorial requirements for the CEGB's AGR Simulators are currently moving away from a straight menu-driven "question-and-answer" session, towards a programmable command list facility whereby a sequence of simulator operations and commands can be specified for each training exercise in turn.

VI. COMPUTER AND INTERFACE HARDWARE

The plant model complexity, together with the display and tutorial facilities requirements, will determine the order of computer size and power necessary for the simulator. The final choice of machines can often wait until some initial studies have been carried out, and it is partly with this in mind that the CEGB's software development strategy has evolved in the way it has.

Plant models are developed and timed for speed of execution on the CEGB's National mainframe computing facility, which gives access to the full range of software development systems. Then, with the knowledge of the relative speeds of other machines on the market (obtained by running standard CEGB simulator benchmark programs), a suitable choice of target machine can be made.

A straight comparison of execution speeds is not, of course, the only criterion - other effects such as cache memory sizes, main memory contention, and machine architecture also make a contribution. However, it is important to

understand that an absolute comparison of running speeds is
not important, as a degree of spare machine capacity should
always be planned into the system to allow for future
upgrades.

The importance of producing standard, machine-transport-
able software is readily apparent in this context, as the
target machine may not be the mainframe on which the software
is developed. Therefore, for maximum transportability, the
target machine should be a standard scientific machine capable
of running standard scientific programs (which will almost
certainly be written in FORTRAN).

The continuing trend towards ever faster standard machines
currently offers no long-term advantage for the use of
special-purpose hardware and software, which may initially
be considered as a viable option to achieve real-time running.
Indeed, the twenty to forty year lifetime of a replica
Power Plant Operation Training Simulator will undoubtedly
include a number of upgrades due to obsolescence of hardware,
thereby further emphasising the importance of transportable
software in order to provide a flexible choice of hardware
suppliers.

The choice of computers for the DPS and display systems
is largely governed by the decision to "simulate" or
"stimulate". If the latter is chosen, as for the Heysham II
AGR Simulator, then the simulator will use hardware which is
as near identical as practicable to that used for the actual
power station systems (i.e. replica hardware), thereby
minimising the manpower resource required to keep the
simulator system in line with on-going releases of the station
software.

With future support and maintenance of hardware and
software in mind, the CEGB's AGR Simulators have all been
developed using standard scientific computers, together with
industry-standard control desk/panel interface hardware.

VII. SOFTWARE DEVELOPMENT STRATEGY

The software for all the CEGB's AGR Simulators has been,
and is being, developed by one group of centrally managed
systems analysts. As an example, the strategy for plant
model software development will be described in detail.

Plant model software codes are all written in the same FORTRAN-based digital simulation language PMSP (Plant Modelling Systems Program [1]), which has been developed and supported within the CEGB over a number of years. As the standard CEGB plant modelling language, it also supports the development of detailed design codes, against which the simulator plant models are compared in their early stages of validation.

PMSP is a highly developed continuous system simulation language, incorporating its own input/output routines, centralised numerical integration, automatic steady state finder and linear analysis routines, together with very convenient debugging facilities. Users also have access to a library of plant model subsystems (e.g. boilers, reactors, pressurisers) which are stored as PMSP "MACRO's" and can be invoked as part of an overall simulation.

In this way, developed and validated codes can be conveniently used by any PMSP user when and where appropriate. PMSP currently runs on a number of mainframe and supermini computers and, being FORTRAN-based, offers a familiar environment for scientific programming staff.

It follows that the adoption of PMSP as the common high level language for CEGB simulator software development, backed up by the use of standard scientific computers, gives a wide range of hardware from which to choose. Also the ability to move staff between projects, particularly in the long-term support role, is considerably eased by this commonality of software systems.

Producing plant model software for all simulators centrally has the advantage that the individual teams in the central group can work closely together, and develop and implement the same techniques. Plant model subsystems are then common to a number of simulators (for nuclear or fossil-fired plant operation training), as are the specific mathematical codes used. An example of the latter is the technique of "Local Eigenvalue Protection" [2], specifically developed by the CEGB for simulator plant modelling to overcome the problems of using fixed-step integration routines in the presence of small time constants.

Long-term support and software enhancement for existing simulators is also carried out under the same central system,

so any problems and their subsequent solutions are readily
made known to, and therefore able to be taken account of by
staff developing software for new simulators. This has
enabled standards to be equated for simulators on different
sites, thus ensuring that details of the features used to
advantage on one simulator are readily available for back-
fitting to all the others.

Broadening the issue to scheduling, Power Plant Operation
Training Simulators must have the ability to communicate
with control desk/panel switches and instruments, and also to
interface to a DPS if this is included. A convenient way
of achieving this is by table-driven software where the
tables are automatically generated from a master schedule or
schedules of the appropriate signals. CEGB power stations
employ such schedules for their actual control and display
systems, and these are set up using the standard CEGB
Information System package INFSYS [3]. This allows easy
entry, editing, sorting and listing of the hundreds of
individual items which make up the schedules (or database).

As these station schedules are accessed via the CEGB's
central mainframe computing facility (on which the simulator
plant model software is initially developed), the equivalent
simulator schedules are conveniently derived from them, also
using INFSYS. In this way, the use of standard CEGB software
database systems takes the onus of their development and
maintenance away from the front line of simulator software
development. Future enhancements of such systems are therefore
automatically available to the simulator teams as "users".

By having a centrally managed organisation for simulator
software development, commonality of commercial and operational
software security standards is readily achievable. The CEGB
has specific procedures covering the provision of computer code
listings and source material (such as datasets residing on mass
storage media), and these apply equally to simulator software
as well as any other. In this way, the commercial confident-
iality of the CEGB's simulator software is protected within
well-defined guidelines.

All code developed on the CEGB's general purpose mainframe computers is operationally protected by standard back-
up procedures. Hence, the AGR Simulator plant model software
(which runs on these machines) is regularly backed up as a
matter of course, without the simulator teams having to carry
out this task themselves.

This all highlights the advantages of adopting standard systems and procedures within a project as large as the development of a Training Simulator, and the CEGB's centrally managed policy for software development consolidates those advantages.

VIII. LONG TERM SUPPORT

The long-term support of simulator hardware is effectively assured by adopting appropriate maintenance procedures and spares holdings. It should always be recognised that a full-scope replica Training Simulator is a major computing installation, and must therefore be run accordingly if a high level of hardware availability is to be achieved.

A similar principle applies to the maintenance and enhancement of the various software systems, and the CEGB's centrally managed approach is fundamental to the way this is carried out. Centralised development and support secures the advantages of commonality of simulator software maintenance procedures, as the people who develop the systems actually maintain them. Also, as previously mentioned, the experience gained from all simulators is more naturally pooled, and therefore benefits each one through a proper programme of back-fitting of those features common to all simulators. This strategy adopted by the CEGB covers not only AGR, but also Conventional Power Plant Training Simulators, as many areas (such as turbine and feed system modelling) are common to both.

On-going support of replica Training Simulators inevitably involves a large number of coding changes over a period of years, in order to reflect actual plant modifications at the power stations concerned, and also to incorporate enhanced training or tutorial facilities as required by the user. It is worth reiterating at this point that the useful lifetime of a replica simulator will be comparable to that of the particular power station - ranging from twenty to forty years, depending upon the type of station involved.

With large simulation codes evolving through many modifications over a period of years, it is essential that formal procedures are observed in each case, from the initial change requests right through to the eventual coding changes themselves.

For the purposes of change control, the documentation

procedure used by the CEGB for simulator software modifications involves three separate forms (as well as the final
annotated coding) - viz:

(a)   Problem Report (PR)

(b)   Modification Request (MR)

(c)   Modification Specification (MS)

A PR is issued by the user if he is not satisfied with
any aspect of the simulator's performance - that is, if a
definite PROBLEM has been identified. In theory, anyone
can issue a PR, but in practice it is the user who normally
has the first-hand experience of most problems.

An MR is also issued by the user, but this is restricted
to requests for any changes to the simulator software which
are required to consolidate or enhance training, either
directly or via the tutorial facilities - that is, if a
MODIFICATION is needed.

Finally, an MS is issued by the software support staff
as soon as a suitable method of implementing the changes
required by a PR or MR has been finalised. Thus, an MS is a
SPECIFICATION for the necessary coding changes which will
themselves be annotated in the software listings, cross-
referencing all the relevant PR, MR and MS numbers.

Whenever a PR, MR, or MS is issued, the originator sends
a copy to each interested party, namely the user and the
relevant software support teams. At an appropriate time
(when either a reasonable number of PRs or MRs have been
issued, or when an urgent change is required), a Modification
Working Group (MWG) meeting is held to discuss these points,
and to assign a priority order for servicing them. This
results in an agreed list of changes to be incorporated into
the next "Simulator Software Release". Such MWG meetings are
therefore held on an ad hoc basis for each particular
simulator, with the user and the software support staff being
the main participants.

It could be argued that such a process can delay the
incorporation of simulator software enhancements, but it
should be remembered that the sheer size and complexity of
the software involved demands such a formal procedure for
change control, otherwise disaster can easily follow.

To put this into perspective, it is often quoted that around two-thirds of the software life cycle is maintenance (and this applies to most software, not just simulators). It is therefore vital to maintain control of such a very complex operation.

Those who believe in, and adopt, a process of "tinkering" and "instant changes" are courting disaster, and this is recognised by all those involved in software development and maintenance who direct their efforts along the lines of proper "Software Engineering" principles.

IX. QUALITY ASSURANCE

The field of Quality Assurance is all-embracing, and governs the practices employed to guarantee the successful outcome of any simulator project. This covers not only the quality of the individual components themselves (e.g. hardware and software), but also the procedures adopted to ensure that project progress is maintained at the necessary level right up to formal acceptance.

It should be appreciated, though, that Quality Assurance does not end here.

After formal acceptance of a simulator, it is equally important that the long-term support and maintenance functions are also subject to well-defined guidelines (Section VIII) in order to maintain quality, and therefore availability and training value, of the simulator.

There is no single best way of managing a particular project - each Organisation has its own methods, and the CEGB is no exception in this respect. As it would take far too long to explain every detail of the running of a CEGB simulator project, a number of major factors will be highlighted instead:

(a) A well-defined project management structure is adopted, covering every aspect of simulator procurement - civil works, hardware and software procurement, together with the user's involvement.

(b) A series of meetings, with membership, is defined in order to monitor and report progress. This ranges from high-level managerial meetings, regular progress meetings

including middle management, to working level meetings
at which the simulator design is advanced.

(c) Quality Assurance of civil works and hardware procurement
is covered by standard CEGB practices adopted for all
related projects, not only simulators. Industry-standard
equipment is used wherever possible, to reduce the risk
of dependence upon a single supplier, and also of having
to test novel systems.

(d) Simulator software is developed and tested according to
specific design procedures. As an example, the validation
of simulator plant model codes is carried out in three
distinct phases:

(i)   In the early development phase, the steady state
       and transient responses of individual simulator
       plant model subsystems are compared with those
       obtained from detailed design or reference codes.
       These latter codes will have been subject to a
       formal process of validation against plant or rig
       tests over a number of years.

(ii)   When the full simulator is operational the tutorial
       staff, who will eventually be using it within
       their training programmes, provide a useful and
       informed input to the overall validation process
       by way of their views regarding the simulator's
       training credibility. It is worth mentioning
       here that about one year should be allocated for
       testing and commissioning a fully integrated
       simulator. If the training need is to be properly
       met, it cannot be done in less time.

(iii) In the case of a replica simulator designed to
       give training before the power station is commis-
       sioned, the final stage of validation is the
       comparison with actual plant responses after
       station commissioning. As it can take a long time
       to obtain this data in a suitable form for a
       meaningful and objective comparison exercise, some
       useful interim views are obtained from the station
       operations staff when they attend revision training
       courses on the simulator.

(e) Previously validated software is stored in a library

of simulator model subsystems, for use on future
simulators and also to be readily available for back-
fitting of enhanced features to existing simulators. In
the case of AGR Simulator plant models, these are held
on a standard CEGB mass storage system which is regularly
backed up according to the central mainframe installation
practices.

(f)   The Design Intent Statement procedure (Section III) serves
      as the central source of all information for the
      simulator, either directly or by reference. In this way,
      for example, the information used in developing the
      original software can be traced right back to its origin
      (such as manufacturer's plant drawings or control system
      design submissions).

(g)   Long-term support and maintenance of the simulator software
      systems is documented by the PR, MR, MS procedures,
      together with suitably annotated coding (Section VIII).
      This then provides a permanent record of all simulator
      software modifications.

X. AGR SIMULATOR SYSTEM DESIGN

Each of the CEGB's AGR power stations is operated within
the concept of a centralised control room. That is, all the
major plant functions necessary for operation of a complete
reactor-boiler-turbine-generator unit (and its auxiliaries)
are carried out by a shift operations team in a central control
room. A typical control room layout would comprise a control
desk ("horseshoe" shaped, for example), and a number of
separate panels, each of which would contain control switches,
meters, lamps, etc.. A large proportion of the information
presented to the operators is in the form of VDU displays
driven by the station Data Processing System (DPS). This
includes alarm messages, and a number of representative plant
"formats" such as a core map of the reactor showing individual
channel gas outlet temperatures.

The AGR Simulators replicate the physical control desk
and panel layouts, together with a simulated DPS, for each
particular power station. A typical AGR Simulator schematic
design is shown in Figure 3, and the most obvious feature of
this arrangement is the fact that the instrument computer
system interfaces to a general purpose central mainframe

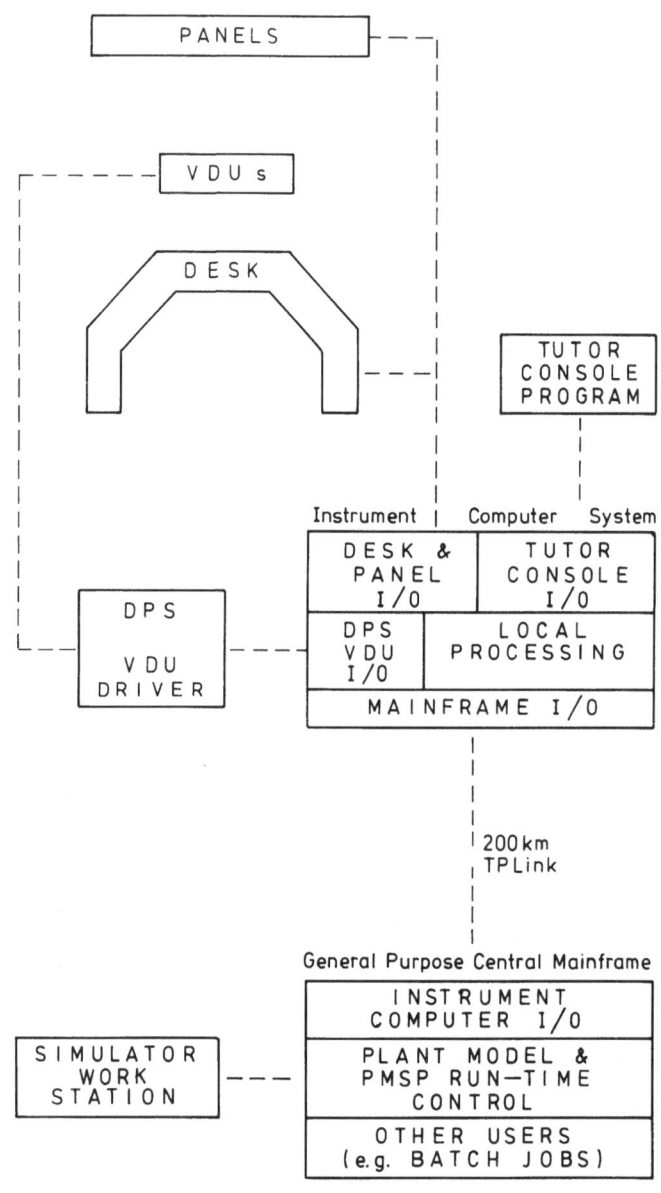

Figure 3

AGR Simulator Schematic Design

machine via a teleprocessing link, over a distance of approximately 200 kilometres. This is probably unique to the
CEGB's AGR Simulators although, in all other respects, the
component parts of the overall system (together with their
interaction) are common to all power plant simulators in
principle. These will be described along with the overall
systems design philosophy, highlighting the main differences
between this and other simulator configurations.

(a) Instrument Computer System

The instrument computer sits at the centre of all
simulator operations. It provides the man-machine interface
with the tutor, and the real-time interfaces with the desk/
panel controls and instrumentation and the plant model computer.
It also represents a simulated plant interface to the DPS,
whenever a fully replicated DPS is used for the simulator.

The machine is usually a standard mini or supermini
computer (PDP 11/44 in the case of the early AGR Simulators,
or VAX 11/780 for the Heysham II AGR Simulator). The major
tasks it performs are described in more detail as follows:

(i) The man-machine interface services the "Tutor Console
     Program" (TCP), and the VDU displays designated as
     "Tutor Formats". The TCP is the means by which the
     tutor controls the simulator and the training session
     from the initial setup phase to real-time operation,
     and contains all the features normally associated with
     this task. These are, for example, mode control
     (initialisation, static test, run, hold, snap, etc),
     fault injection, tutor override of control switches, and
     selection of particular tutor formats for display on
     particular VDU tubes. In the latter case, the
     instrument computer sets up the VDU displays via the
     VDU driver, which may be regarded as the equivalent of
     a number of intelligent display terminals.

     These tasks are obviously a standard requirement, in
     one form or another, on all comparable Training Simulators.

(ii) Another standard simulator task performed by the
     instrument computer is the real-time interface with the
     desk/panel controls and instrumentation. This handles
     all the analogue and digital inputs and outputs
     associated with switches, push-buttons, lamps, meters, etc.

The interface hardware used on all the CEGB's AGR
Simulators is the CAMAC system, which is a non-proprietary
international standard.

(iii) After scaling, the desk/panel inputs and outputs provide
the real-time interface link variables with the plant
model computer. In most simulators, this may be a
number of other processors incorporating DMA or shared-
memory links with the instrument computer, but the CEGB's
AGR Simulators use a remotely-located general purpose
mainframe to run the total plant model in real time. The
two machines therefore communicate via a high speed
teleprocessing link in this case. Not all the required
input/output variables are generated by the total plant
model, as the training requirements do not necessarily
demand that all plant is simulated (e.g. only two, four
or nine boilers, instead of the actual twenty four: or
a single thermocouple, where the actual plant may use a
number for redundancy). Consequently, the instrument
computer generates some of those variables which either
do not have a fully "closed-loop" interaction with the
main plant model (such as turbine supervisory calculations),
or which only require a "fan-out" from key variables
(e.g. multiple thermocouples). This comes under the
general heading of "Local Processing".

(b)   General Purpose Central Mainframe

A specific feature of the CEGB's AGR Simulators is their
use of a general purpose mainframe for running the real-time
total plant model. In order to fully appreciate the
significance of this, it is worthwhile understanding the
provision of computing services in the CEGB.

The CEGB is a publicly-owned utility which provides in-
house software development and computing bureau services.
Engineering systems services are supported via a large mainframe installation in London, providing both interactive and
batch processing facilities which are networked to user
terminals. Virtually all major studies involving plant modelling use this service, so there is a large investment in
mainframe computer programs covering all aspects of the CEGB's
design and operational work. The standard CEGB simulation
language for dynamic plant modelling is PMSP (Section VII and
[1]), which is held and supported at this installation, as
are a number of other packages of general use. The machines
which provide these facilities are two IBM-compatible main-

frames, each one being rated at over 14 Mips (Millions of instructions per second).

When the initial feasibility studies were carried out for running total plant models in real time, the CEGB's design codes were used as benchmarks. It was found that the complexity of the modelling requirements for the early AGR Simulators could only be realistically met by using a machine of the power of the mainframes.

Also, as the CEGB's design codes were conveniently accessible, together with PMSP, the mainframe was deemed to be the appropriate system on which to run the simulator total plant models. By using the general purpose computers, the simulator plant models had to run in the standard operating system environment, which involved batch or interactive services within the framework of a virtual machine. On the face of it, this may seem incompatible with the real-time running require-ment of a Training Simulator, as batch and general interactive work are non real-time applications, whilst the paging and swapping activities of a virtual machine are definitely not compatible with real-time operation.

However, these apparent problems were eliminated by writing a special application program which enabled the simulator plant model to be run as a very high priority batch job, capable of being controlled interactively during execution. The paging process was overcome by specifying that the job must occupy "real" and not "virtual" core.

This application program allows each simulator plant model to be controlled from a standard keyboard/VDU terminal, which can be regarded as a simulator work station for the purposes of development and off-line testing. All the standard simulator functions are available by this means, so the work station is a direct substitute for the desk/panels and instrument computer, as far as plant model input/output and simulator mode control are concerned. It can also be used for remote monitoring of the simulators from London, where the plant model software development teams are located.

In this way, the simulator plant models can run in real time alongside other batch and interactive applications on the general purpose mainframe system. The power of the system is such that two simulators can run concurrently in real time on the same mainframe processor, and still enable other jobs to share the machine.

(c)   200 km Teleprocessing Link

   The instrument computer communicates with the total plant
model running on the general purpose mainframe by means of a
high speed teleprocessing link. For each of the early AGR
Simulators, this comprises a half-duplex bisynchronous configuration using two standard British Telecom Tariff "T" telephone
lines, each having a bandwidth of 9.6 kB (kilo-Baud). The
simulator incorporates line driver tasks at each end, which
direct blocks of information to the "next free line". In
this way, if one line is unavailable (perhaps due to a
transient fault), then the simulator will continue to run on
the remaining line. This might cause a temporary reduction
in the update rate of desk/panel input/output, but the system
is self-correcting when the second line becomes available again.

   The Heysham II AGR Simulator is also designed to use a
teleprocessing link but, due to the increased amount of real-
time information which must be transmitted, it will use a single
line with a bandwidth of 2 MB (i.e. one hundred times greater
than for the earlier simulators) within the British Telecom
Megastream Network. The computers at each end of this line
will interface with it via NSC HYPERbus hardware, which is
capable of full-duplex operation.

   The teleprocessing link between the instrument computer
and the mainframe is used for the initial setup phase and for
mode control, as well as for the real-time interchange of
variables relating to the desk/panel and DPS (because all these
events affect the plant model). For this reason, the communications software is organised in two distinct modes -
termed CONVERSATIONAL and REAL:

(i)   CONVERSATIONAL mode is used for setting up, selecting a
      particular set of starting conditions, performing a static
      test (or "discrepancy" check) to ensure that the desk/
      panel control switches and other inputs are in the correct
      state, and then initialising the simulator ready for real-
      time running. These operations involve the instrument
      computer transmitting tutor commands to the mainframe,
      and receiving in turn the required desk/panel inputs
      corresponding to the selected starting conditions. This
      information is acknowledged on receipt, and retransmitted
      if necessary (i.e. a "handshaking" operation between the
      two computers).

(ii) When the setup phase is complete, the lines are selected
      to REAL mode ready for real-time running. The instrument
      computer sends the desk/panel switch states in blocks of
      1, 2, 4 and 8-bit integers to the mainframe, as inputs to
      the plant model. These are termed DCVARs (Digital
      Control VARiables). The Heysham II AGR Simulator also
      uses 32-bit RCVARs (Real Control VARiables) to handle
      analogue values mainly set by the tutor (e.g. to control
      isolating valve lifts, or to select the size of a boiler
      tube leak).

      Interleaved with these blocks being sent by the instrument
      computer are the required outputs from the plant model,
      which are being transmitted from the mainframe for
      eventual display on the desk/panel and DPS. These are
      sent in blocks of 16-bit scaled fixed point quantities
      (in the case of the early simulators), and as 32-bit
      floating point values for the Heysham II AGR Simulator.
      They are split into two groups, depending upon whether
      a fast update rate is needed (e.g. for desk/panel
      analogue instruments), or whether a slow rate is sufficient
      as in the case of VDU formats which may only be updated
      every few seconds. These variables are termed MOVFST
      (Model Output Variable - FaST) and MOVSLW (Model Output
      Variable - SLoW). No acknowledgement of receipt of these
      blocks is issued by either computer in REAL mode, and
      consequently no retransmission takes place, because the
      information would be out of date by the time it was sent
      again and finally acknowledged. It is better in this
      case to lose a block completely rather than suffer the
      overhead of retransmission.

      In over four years of virtually continual training use,
the AGR Simulators and their teleprocessing links have proved
to be very reliable in operation. The fairly large geographical separation between the instrument computer system
and the mainframe has had no adverse effects on the reliability
of the overall system. Indeed, the only difference between
this arrangement and a totally dedicated local system is the
teleprocessing link itself.

XI. CONCLUSIONS

      The relative merits of the CEGB's AGR Simulator design,
with respect to using the general purpose mainframe system,
can be briefly summarised as follows:

(a)   Standard CEGB software development packages, such as PMSP
      and INFSYS are readily accessible and usable.

(b)   The whole of the main plant model can be executed in a
      single processor, thereby eliminating the task of having
      to split up the code into parallel processes (and
      therefore the need to ensure correct synchronisation
      between those processes).

(c)   The development and target machines are the same.
      Therefore, no transfer of code is necessary to produce
      the final production versions of the simulator plant
      model software. Also, on-going enhancement of the plant
      models is carried out without the need for either a
      separate development machine or scheduling of the
      simulator target machine (which might otherwise conflict
      with training courses).

(d)   Progressive upgrades or replacement of the general
      purpose mainframes are entirely transparent to the
      simulator application. As an example, the AGR Simulators
      have run on five different IBM-compatible mainframes since
      their inception, without any software modifications being
      necessary.

(e)   When the simulators are not being used for training
      courses, the plant model computer (i.e. the mainframe)
      is not left idle, as it forms part of the CEGB's central
      computing service.

      The CEGB's design and procurement strategy for its AGR
Training Simulators has proved very successful. The first
three simulators were formally commissioned during the period
from May 1981 to April 1983, and two of them provided
operational training prior to commissioning of the actual
power stations. Since then, all three have been in regular
use, forming an integral part of the CEGB's AGR Operator
Training Programme. This ranges from "generic" use to
demonstrate the physical and operational details of individual
plant items, right through to full revision training for
experienced plant operations shift teams.

      In conclusion, the "general purpose mainframe" approach
for running the plant model software in real time has proved
to be a practical alternative to totally dedicated systems,
and indeed has some real advantages to offer.

ACKNOWLEDGEMENTS

The author would like to thank Mr J. J. Williamson, Director of CEGB Computing and Information Systems Department, for his permission to publish this article. Thanks are also due to colleagues at CEGB HQ Computing Centre for useful comments and discussions during its preparation.

REFERENCES

[1] Laderman L.M. and Richards C.J.
     'The Plant Modelling System Program Reference Manual',
     CEGB HQ Computing Centre Report CC/N755.

[2] Budd G.C.
     'Local Eigenvalue Protection in Power Plant Modelling
     for Training Simulators',
     Paper presented to I.N.E. Conference (Cambridge,
     England) on "Simulation for Nuclear Reactor Technology" -
     April, 1984.

[3] Bazley D.E.
     'An Introduction to INFSYS',
     CEGB HQ Computing Centre Report CC/N728.

References 1 and 3 are CEGB reports. Any requests for copies thereof should be addressed to : The Librarian, CEGB HQ Computing Centre, 85 Park Street, London SE1 9DY, U.K.

PARALLEL PROCESSING FOR NUCLEAR SAFETY SIMULATION

A.Y. Allidina and M.G. Singh

Control Systems Centre, University of Manchester
Institute of Science and Technology, U.K.

B. Daniels

Systems Reliability Service
United Kingdom Atomic Energy Authority
Culcheth, U.K.

I. INTRODUCTION

With the advances made in computer technology, there
has been much interest in the use of this technology for
System Design, Reliability and Safety aspects. One general
application area is in the nuclear industry. A particular
problem in the nuclear industry (and other industries) is to
simulate system models sufficiently fast in order to, for
example, provide real-time predictive information to plant
operators. However, in order that the models might represent
the corresponding plants in a realistic way, these models are
invariably complex. This makes simulation rather difficult
and very often much slower than real-time if standard solution
techniques are used together with modest present-day computer
technology. Given this problem, it is clearly desirable to
investigate ways of improving the simulation speed without
dramatic increases in cost. In this chapter, we are concerned
with the above objective. The eventual aim of the research
work reported here is to investigate the application of the
developed techniques to the Nuclear code RELAPV [Allidina
(Editor) 1984].

One way of improving the simulation speed is to make use
of parallel computing facilities in order that the required
computation may be done in parallel. A typical simulation
problem consists of having to solve differential equations,
partial or ordinary. In order to obtain a solution, it is
necessary to discretise the differential equations. This
step introduces an approximation, the quality of which

depends on the type of discretisation used. If the discretised equations are to be solved using parallel computing
facilities, then the ease with which the calculations can be
arranged in order to be performed in parallel also depends
on the type of discretisation used. The two objectives of
a discretisation scheme of achieving stability and good
accuracy, and enabling parallel computing to be applied with
ease, can often be contradictory. One can attempt to enhance
parallelism either after discretisation or before discretisation of a set of differential equations. For example,
there are some parallel methods for the numerical solution
of ordinary differential equations in which the discretisation
is done so as to provide for parallelism, i.e. to develop
a 'parallel front' of computation (Miranker and Liniger (1967),
Worland (1976), Franklin (1978), Katz et al (1977)). In this
chapter, however, we discuss a method amenable to parallel
computing after the discretisation has been done in order
not to interfere with the well established discretisation
schemes so that stability and accuracy are maintained.

The issue of parallelism can be addressed at many
different levels (Burks 1981), Schendel (1981)) with
relation to different computer architectures and to concepts
in parallel numerical methods. As far as architecture is
concerned one can think of parallelism at the instruction
execution level (e.g. 'pipe-lining'), or of vector and array
processors executing a stream of single instructions with
multiple data (SIMD architecture), or of special multiple
instruction architectures (multiprocessing architectures)
with several streams of instructions being executed simul-
taneously (MIMD) (Burks(1981)). One can have multiprocessor or multicomputer networks having different
configurations (e.g. common multi-bus systems (Arnold et
al (1983)), or systems having a ring structure (Brasch et
al (1981)).

As far as concepts in parallel numerical mathematics
are concerned, it is possible to consider parallel execution
of elemental operations when evaluating mathematical
expressions (e.g. when evaluating expressions like 'Horner
scheme' with 'Log-sum-algorithm'). It is also possible to
consider, at a higher level, inherent parallel operations
in well established algorithms (Schendel (1981)). Finally,
at a still higher level, one can specify larger tasks for
separate computers to be executed in parallel while using
decomposition or decomposition-coordination techniques (e.g.

See Travassos and Kaufman 1980, Singh et al.1983). The
execution of these tasks can be done in a synchronous or
asynchronous mode. It is this last issue of parallelism at
a rather high level that we are concerned with. This is
related to the use of a multicomputer network rather than an
array processor, for example. Thus the approach taken is to
split the overall problem into a number of interacting sub-
problems using hierarchical techniques of decomposition-coordination. The subproblems can then be solved on different
processors (in parallel) while the interactions between the
sub-problems are taken care of by 'coordinating' the solution
of the sub-problems.

In view of the above discussion, in this chapter we
discuss a particular decomposition-coordination technique
which enables tasks to be performed in parallel when solving
large sets of equations of a specific structure. This
structure arises, for example, when solving a set of partial
differential equations using particular types of discretisation schemes. In such an application it is necessary to
solve the resulting equations at each time level, and therefore the developed method, 'hierarchical system solver' needs
to be used many times.

As stated earlier, the prime objective for the development
of the techniques is to investigate the possibility of
speeding up the Nuclear code RELAPV. However, in this chapter
we consider a scaled-down problem, while maintaining the
same discretisation scheme as that used in RELAPV.

In order to show generality of the developed method in
terms of applicability to other discretisation schemes, the
method is also applied to solve a set of partial-differential-
equations (PDEs), describing gas flow through a pipeline,
when an implicit discretisation scheme is used. This results
in a set of non-linear equations to be solved at each time
level.

The chapter is organised as follows. In Section II, the
problem is defined and a basic decomposition-coordination
technique is used to obtain the solution by specifying
independent tasks. In Section III, the method is applied to
solve a set of PDEs which are discretised implicitly, resulting in non-linear equations, while in Section IV, it is shown
how the method may be applied to the simplified reactor model,
mentioned above, which is discretised in the same way as RELAPV.
Finally, some concluding remarks are given in Section V.

II. PROBLEM DEFINITION AND SOLUTION METHODOLOGY

A simulation problem may consist of solving a set of algebraic equations and/or ordinary-differential-equations and/or partial-differential equations. After time and space discretisation where appropriate, the problem amounts to solving at each time level, for example, a set of equations of the form:

$$F(z) = 0 \tag{1}$$

where $z \in R^{n_z}$ and $F: R^{n_z} \to R^{n_z}$. We consider the case when eqn. (1) can be put into the following form:

$$F_i(x_i, y_i) = 0, \quad i=1,2,\ldots.\nu \tag{2}$$

and

$$y_i = A_i x_{i-1} + B_i x_{i+1}, \quad 1=1,2\ldots.\nu \tag{3}$$

where

$$x_i \in R^{n_{x_i}}, \ y_i \in R^{n_{y_i}}, \ F_i : R^{n_{x_i}} \times R^{n_{y_i}} \to R^{n_{x_i}},$$

$$A_1 = 0 \text{ and } B_\nu = 0.$$

It will be shown later in Sections III and IV how such equations can arise when solving a set of partial differential equations.

We now consider ways of solving the above problem with the use of a parallel computing facility. One possible way of doing this would be to use successive linearisation, that is, the equations can be linearised around a certain point and the resulting system of linear equations solved, after which the linearisation is repeated, and so on. In such a case, in order to solve the set of linear equations, one can use suitable 'parallel-system-solvers'. There is a vast amount of literature on this subject, see for example, Barlow and Evans (1982), Evans and Haghighi (1982) and Halada (1981). Another approach to solve the set of eqns. (2) and (3) is to use decomposition-coordination schemes. This involves decomposing the set of partititioned eqns. (2) and (3) into independent parallel problems by introducing suitable coordination variables. This approach goes along the lines of decomposition in large-scale optimisation problems (Himmelblau 1973, Lasdon 1970, Wismer 1971, Singh and

Titli 1978, Findeisen et al 1980). It is possible to use well-established decomposing-coordination techniques after adapting them to solve the system of eqns. (2) and (3). It is necessary to use efficient coordination strategies in order that the solution of the sub-problems does not have to be repeated many times. In this chapter we are concerned with the latter approach of using decomposition-coordination.

One of the basic decomposition-coordination methods is the 'direct' method (Findeisen 1980). In our case the procedure consists of treating the variables y_i, i=1,2....ν, as the coordinating variables (hence the name 'direct'). For given y_i, i=1,2....ν it is possible to define ν independent local problems to be solved in parallel. The overall procedure is as follows (Malinowski et al 1984, Allidina et al 1985):

(i) Local problem i, i=1,2,....ν.

For given y_i, solve $F_i(x_i,y_i) = 0$ for $\bar{x}_i(y_i)$.

(ii) Coordinator problem

Find $\hat{y} = (\hat{y}_1,\hat{y}_2....\hat{y}_\nu)$ such that

$$\hat{y}_i = A_i\bar{x}_{i-1}(\hat{y}_{i-1}) + B_i\bar{x}_{i+1}(\hat{y}_{i+1}), \quad i=1,2....\nu. \qquad (4)$$

Figure 1 shows the overall scheme.

As far as the local problems are concerned, it is assumed that, given y_i, a solution $\bar{x}_i(y_i)$ exists for $F_i(x_i,y_i) = 0$. The local problems may be solved in different ways depending on the particular equation set $F_i(x_i,y_i) = 0$. For example, the Newton-Raphson procedure may be used:

$$x_i^{N+1} = x_i^N - \left[\frac{\partial F_i}{\partial x_i}\bigg|_{x_i=x_i^N}\right]^{-1} F_i(x_i^N,y_i). \qquad (5)$$

Starting from an initial guess x_i^o, eqn. (5) is used successively for N=1,2.... until convergence is achieved. It is, of course, assumed that F_i is differentiable and that the Jacobian $\partial(F_1,F_2....)/\partial(x_1,x_2....)$ is non-singular. The

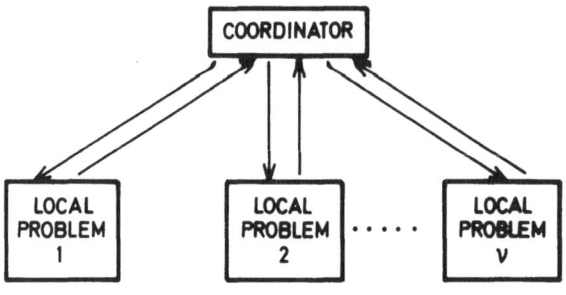

Figure 1

A Hierarchical Scheme

convergence conditions for the above scheme have been extensively examined and are well known.

Since the solutions $\bar{x}_i(.)$ of the local problems are
usually not available in explicit form, the coordinator
problem must be solved iteratively. This implies that for
each iteration at the coordination level resulting in new
values for the variables y_i, the local problems must be
solved again. Therefore it is of crucial importance that
the strategy used for solving the coordinator problem is as
efficient as possible. We will consider two coordination
strategies. In order to make the notation more convenient,
the coordinating conditions (4) can be written as:

$$y = C\,\bar{x}(y),\qquad\qquad\qquad (6)$$

where
$$y = (y_1, y_2, \ldots y_\nu),\ \bar{x}(y) = (\bar{x}_1(y_1), \bar{x}_2(y_2), \ldots \bar{x}_\nu(y_\nu))$$

$$\text{and } C = \begin{pmatrix} 0 & B_1 & & & & \underset{\sim}{0} \\ A_2 & 0 & B_2 & & & \\ & A_3 & 0 & B_3 & & \\ & & & A_{\nu-1} & 0 & B_{\nu-1} \\ \underset{\sim}{0} & & & & A_\nu & 0 \end{pmatrix}$$

Coordination strategy 1 - Relaxation

The coordinating process has to be initiated with an initial guess y°. In the partial-differential-equation example considered, the initial guess y° at the new time level could be taken as the value of $\hat{y}$ at the previous time level. Then at any iteration step M (M > 0), the local problems are solved for $\bar{x}(y^M)$. The coordinator then corrects the value of y according to:

$$y^{M+1} = (1-\omega) \ y^M + \omega \bar{y}^{-M+1}, \qquad (7)$$

where $\quad \bar{y}^{-M+1} = C\bar{x}(y^M) \qquad (8)$

and $0 < \omega < 1$ is the relaxation parameter to be chosen so as to provide for the best convergence possible. In particular, for $\omega=1$, we obtain a re-injection (or Jacobi-type) iterative strategy. This coordination strategy is rather simple with the advantage that a small amount of information needs to be passed between the coordinator and the sub-systems. However the disadvantage is that many iterations may be required.

Coordination strategy 2 - Newton-Raphson

Another strategy may be used for solving the coordinator problem of eqn. (6). Let us define:

$$D(y) \triangleq y - C\bar{x}(y). \qquad (9)$$

Assuming that D is a differentiable function of y and $\dfrac{\partial D}{\partial y}$ is non-singular, then at any iteration step, the coordinator corrects the value of y according to:

$$y^{M+1} = y^M - \left(\left. \frac{\partial D}{\partial y} \right|_{y=y^M} \right)^{-1} D(y^M), \qquad (10)$$

where $\quad \dfrac{\partial D}{\partial y} = I - C\dfrac{\partial \bar{x}}{\partial y} .$

is an expression of the Jacobian of D with the vector y.

The differentiability of D is implied by that of $\bar{x}(\cdot)$ which can be attained under reasonable and well known conditions (implicit function theorem). In order to compute the Jacobian, it is necessary to find

$$\left[\frac{\partial \bar{x}}{\partial y}\right] = \text{block-diag} \left[\frac{\partial \bar{x}_i (y_i)}{\partial y_i,}\right] . \tag{12}$$

$\dfrac{\partial \bar{x}_i (y_i)}{\partial y_i,}$ can be obtained in parallel at the sub-system level

after the local problems have been solved. By differentiating eqn. (2) it can be seen that:

$$\frac{\partial \bar{x}_i}{\partial y_i} = - \left[\frac{\partial F_i}{\partial x_i,}\bigg|_{\bar{x}_i(y_i),y_i}\right]^{-1} \frac{\partial F_i}{\partial y_i,}\bigg|_{\bar{x}_i(y_i),y_i} . \tag{13}$$

The inverse of $\left[\dfrac{\partial F_i}{\partial x_i,}\right]$ required in the above equation is

in fact available from the last iteration at the sub-system level (see eqn. (5)).

It should be noted that it is not necessary to transfer the entire matrices $\dfrac{\partial \bar{x}_i (y_i)}{\partial y_i,}$ from the sub-systems to the

coordinator, but only the appropriate part of the product $C\dfrac{\partial \bar{x}}{\partial y}$ from each sub-system. Since the matrix C is sparse,

this would result in considerable saving in communication time between the sub-systems and the coordinator.

A further point of interest is that if the functions F_i are linear, then the above procedure requires only one iteration at the coordination level. The proposed scheme can then be considered as a particular parallel linear-solver. The discretisation scheme used in RELAPV, for example, results in a set of linear equations to be solved at each time-level. The application of the proposed method to the simplified reactor model with a RELAPV type of discretisation will be discussed in Section IV. First, however, the method is applied to solve a set of non-linear equations resulting from the use of an implicit scheme on the set of PDEs describing gas-flow in a pipeline.

III. SOLUTION OF GAS-FLOW EQUATIONS USING AN IMPLICIT DISCRETISATION

The three non-linear partial differential equations which define one-dimensional flow of gas in a pipeline are derived by considering conservation of mass, momentum and energy through an element of pipe. These equations are:

$$\rho_t + \rho v_\ell + v \rho_\ell = 0, \tag{14}$$

$$v_t + v v_\ell + zR\left[T_\ell + (T/\rho)\rho_\ell \right] + \tfrac{1}{2}\frac{f}{D} v|v| + g\sin\Theta = 0, \tag{15}$$

$$T_t + vT_\ell + (zRT/b)v_\ell - (c\rho/b)v_\ell - \tfrac{1}{2}\frac{f}{D} b|v^3| + 4\lambda(T-T_o)/D\rho b = 0. \tag{16}$$

The subscripts 't' and 'ℓ' denote partial derivatives with respect to time and space respectively.

The variables are:

ρ - density
v - speed
T - temperature

The constants are:

z - compressibility factor.
R - gas constant.
f - friction factor defined by Darcy-Weisbach relation, for the shear stress, $\frac{f}{8}\rho v|v|$, at the wall.
D - pipe diameter.
g - acceleration due to gravity.
Θ - angle of inclination (to the horizontal) of the pipe.
$\left.\begin{array}{c}b\\c\end{array}\right\}$- constants in assumed relationship $H = a+bT+c\rho +P/\rho$, where H is the enthalpy and P the pressure.
λ - thermal conductivity of pipe.
T_o - outside temperature.

To completely define the problem, initial conditions for all the variables and a suitable set of boundary conditions, must be specified. For this example ρ, v and T are specified at t_o (initial time) along the pipe, and ρ and T are specified upstream and v downstream for all time in the interval $[t_o, t_f]$, where t_f is the final time.

Let the discrete time and space steps be τ and h respectively. Let $var_{i,j}$ denote the value of the variable (ρ, v or T) at the grid point whose space coordinate is ih and time coordinate is $j\tau$. The value and the derivative of the variable at the centre of a grid may be approximated by:

$$var_{i+\frac{1}{2},j+\frac{1}{2}} = \tfrac{1}{4}(var_{i,j} + var_{i+1,j} + var_{i+1,j+1} + var_{i,j+1}) \underline{\underline{\Delta}} |\Delta| var,$$
(17)

$$\partial/\partial t(var_{i+\frac{1}{2},j+\frac{1}{2}}) = \frac{1}{2\tau}(var_{i,j+1} - var_{i,j} + var_{i+1,j+1} - var_{i+1,j})$$

$$|\underline{\underline{\Delta}}| \; \Delta \; (var)/2\tau,$$
(18)

and

$$\partial/\partial x(var_{i+\frac{1}{2},j+\frac{1}{2}}) = \frac{1}{2h}(var_{i+1,j+1} - var_{i,j+1} + var_{i+1,j} - var_{i,j})$$

$$|\underline{\underline{\Delta}}| \; \Gamma \; (var)/2h.$$
(19)

Thus the values and derivatives of the variables at the grid centres are expressed as functions of the values at the four adjacent grid points. Applying this discretisation scheme to eqns. (14), (15) and (16) using K grids in the space discretion yields 3K non-linear algebraic equations for each time level. These are:

$$\underline{F}_1 = \Delta(\rho) + \alpha v \Gamma(\rho) + \alpha \rho \Gamma(v) = 0,$$
(20)

$$\underline{F}_2 = \Delta(T) + \alpha v \Gamma(T) + (\beta T \Gamma(v))/b + \phi \rho \Gamma(v) - \gamma |v^3|/b +$$

$$\eta(T-T_o)/\rho = 0,$$
(21)

and

$$\underline{F}_3 = \Delta(v) + \alpha v \Gamma(v) + \beta \Gamma(T) + (\beta T \Gamma(\rho))/\rho + \gamma |v|v + \mu = 0$$

(22)

where the operators Δ and Γ are defined in eqns. (18) and (19) and the constants are:

$\alpha = \tau/h$
$\beta = zR\alpha$
$\phi = -c\alpha/b$
$\gamma = f\tau/D$
$\eta = 8\lambda\tau/Db.$

[To ensure stability the Courant criterion $\tau < h/(|v|+s)_{MAX}$ must be satisfied where s is the speed of sound].

    This discretisation scheme is implicit and gives rise to a set of equations which have to be solved iteratively. It is worth pointing out that for the case of the pipeline equations, the method of characteristics is a better simulation scheme since the characteristic lines are easily obtained and simulation can be carried out very rapidly. However, since the derivation of the characteristic lines is not easy, both for more complex systems and for non-hyperbolic systems, it is not a suitable method for the testing of general decomposition and coordination methods. Also, by virtue of being an explicit scheme, decomposition is trivial.

    The problem of solving eqns. (20) - (22) amounts to solving a set of equations of the form given by eqn. (1), with $n_z = 3K$. Given initial conditions $\rho_{x,o}$, $T_{k,o}$ and $v_{k,o}$, k=0, 1....K and boundary conditions $\rho_{o,j}$, $T_{o,j}$, $v_{K,v}$, j=1,2....,M

(M is total number of time steps), the unknowns at time level j+1 are:

$$[\rho_{1,j+1}\cdots,\rho_{K,j+1},\ v_{o,j+1}\cdots,v_{K-1,j+1},T_{1,j+1}\cdots T_{K,j+1}].$$

The entire set of equations can be solved without any decomposition using Newton-Raphson. This will provide a basis for comparison for the decomposition-coordination method.

    Now, as far as decomposition is concerned, it corresponds to splitting the pipe into a number (v) of 'sub-pipes' and solving the simulation problem associated with each sub-pipe independently (and in parallel on a parallel computing facility). The task of the coordinator is to ensure that the solution obtained is the same as if an un-decomposed method were applied.

We consider the case when eqn. (1) can be put into the form as given by eqns. (2) and (3). For this example the 'coupling parameters', y_i are chosen to be compatible with the boundary conditions. Choosing space points $K_1, K_2, \ldots, K_\nu$ with $K_\nu = K$ (see Fig. 2), we have:

$$y_1 \triangleq v^2_{K_1,j+1} = B_1 x_2,$$

$$x_1 = [\rho_{1,j+1}, \ldots, \rho^1_{K_1,j+1}, v_{0,j+1}, \ldots, v_{K_1-1,j+1}, T_{1,j+1}, \ldots,$$
$$T^1_{K_1,j+1}] ,$$

$$y_i \triangleq \left[\begin{array}{c} \rho^1_{i-1,j+1} \\ T^1_{K_{i-1},j+1} \\ \text{--------} \\ v^2_{K_i,j+1} \end{array}\right] \begin{array}{l} \left.\begin{array}{c} \\ \\ \end{array}\right\} \text{elements of } x_{i-1} \\ \\ \left.\begin{array}{c} \\ \end{array}\right\} \text{elements of } x_{i+1} \end{array} \quad = A_i x_{i-1} + B_i x_{i+1}$$

$$x_i \triangleq [\rho_{K_{i-1}+1,j+1}, \ldots, \rho^1_{K_i,j+1}, v^2_{K_{i-1},j+1}, T_{K_{i-1}+1,j+1}, \ldots,$$
contd/d. $\quad T^1_{K_i,j+1}]^T ,$

$$y \triangleq \left[\begin{array}{c} \rho^1_{K_{\nu-1},j+1} \\ \\ T^1_{K_{\nu-1},j+1} \end{array}\right] \quad = A_\nu x_{\nu-1} ,$$

$$x_\nu \triangleq [\rho_{K_{\nu-1}+1,j+1}, \ldots, \rho_{K_\nu,j+1}, v^2_{K_{\nu-1},j+1}, \ldots, v_{K_\nu-1,j+1}, T_{K_{\nu-1}+1,}$$
contd/d. $\quad _{j+1}, \ldots, T_{K_\nu,j+1}]^T .$

With respect to eqn. (3), the A_i matrices therefore contain two unity elements, the rest being zero, whilst the B_i matrices contain one unity element, the rest being zero.

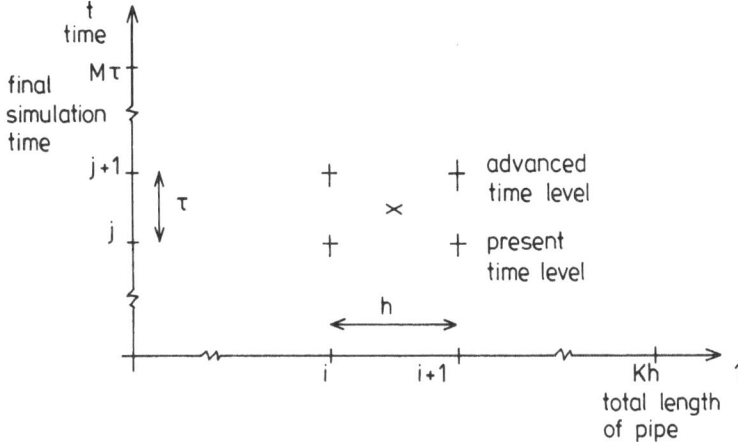

Figure 2

Discretisation in the {1,t} plane

        The number of sub-systems was set to three with a
variable number of grid points, and both coordination
strategies (relaxation and Newton-Raphson) were tried. The
local problems were solved using Newton-Raphson. The
optimal value of the relaxation-parameter ω was found to be
0.5.   With ω = 0.5 the average number of iterations between
the coordinator and the sub-systems was around 16, while
with the Newton-Raphson coordinating strategy, the average
number was 3.   Figure 3 gives a comparison of the three
cases: (i) no decomposition, (ii) decomposition with relaxation as the coordination strategy, and (iii) decomposition
with Newton-Raphson as the coordination strategy. The same
set of initial and boundary conditions, and error tolerance
was used for all the three cases, and the total simulation
horizon was 1000 seconds. The results are for a single
computer, so the decomposed method would be approximately
three times faster than shown in Figure 3 if run on a multi-
computer network.

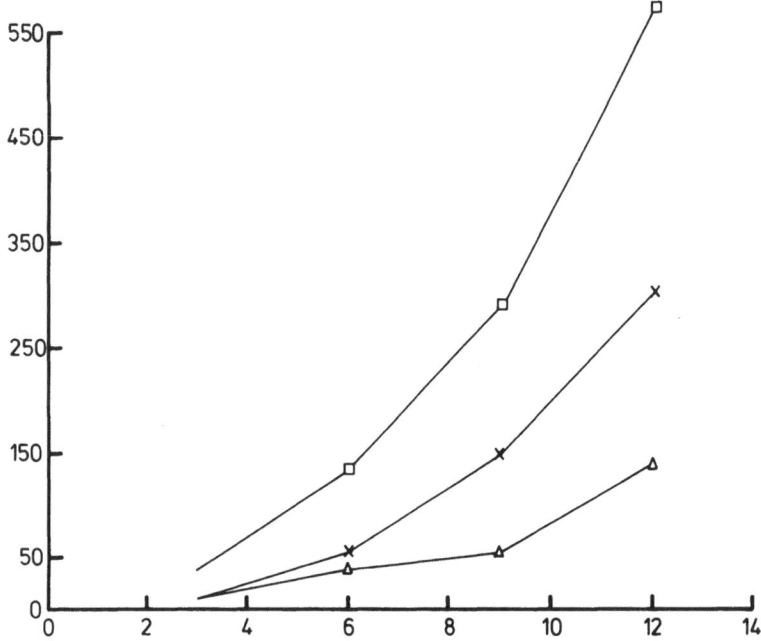

x   Implicit - no decomposition   ☐   Direct - relaxation
                                                strategy
Δ   Direct - Newton-Raphson

Figure 3

- C.P.U. Time against total number
of grids in ℓ-direction

        The example described above shows the potential of the
method given in Section II. In this example, the discretisation used resulted in non-linear equations to be solved at
each time level. The primary aim of the work reported here
as stated before, is to investigate the nuclear code RELAPV.
The development of this code has led to the use of a special
discretisation scheme, which results in a set of linear
equations to be solved. In the next Section, we discuss the
application of the decomposition-coordination method to the

simulation of a simplified reactor model, which is discretised in the same way as RELAPV.

IV. APPLICATION TO RELAPV TYPE OF DISCRETISATION

As stated in the introduction, the main objective for
the development of the decomposition-coordination techniques
is to investigate the possibility of speeding up the nuclear
code RELAPV. However, in this chapter, we only consider a
simplified reactor model together with the discretisation
used in RELAPV. It is briefly shown how the decomposition-
coordination method may be used for this problem.

Simplified Reactor Model

Figure 4 shows a schematic diagram of the reactor
system to be considered. The basic features are a Fluid
Model, a Heat Exchanger Model and a Core Model.

Fluid Model. The fluid is used in transporting the heat from
the core to the heat exchanger and in moderating the neutrons.
In the reduced model, the prime simplification is that the
flow is always single phase. As long as the system pressure
is high enough, the simplification is correct. This obviates
the need for manipulating steam tables. However, this
constraint results in a reduction in the possible reactor
transients that can be simulated and the effect of this on the
decomposition techniques will be unknown. In addition,
exploitation of steam data manipulation techniques cannot be
tried. A further simplification is that flow is assumed to be
laminar throughout, and the pipe contains no valves, T-junctions
and area changes. The fluid is driven around the closed
circuit by a pump which has linear characteristics.

Present work, however, involves including two-phase flow,
changing pipe area and turbulent flow. Provided transients
are small, only a small part of the steam table would need to
be incorporated. (The transient simulated is pump coastdown).

Core Model. At present no model for the neutron-kinetics has
been incorporated but instead a constant heat generation rate
is assumed. The value taken is assumed constant throughout
all the nodes of the heat structure. The rate of heat
production is equivalent to that of one fuel assembly in a PWR.

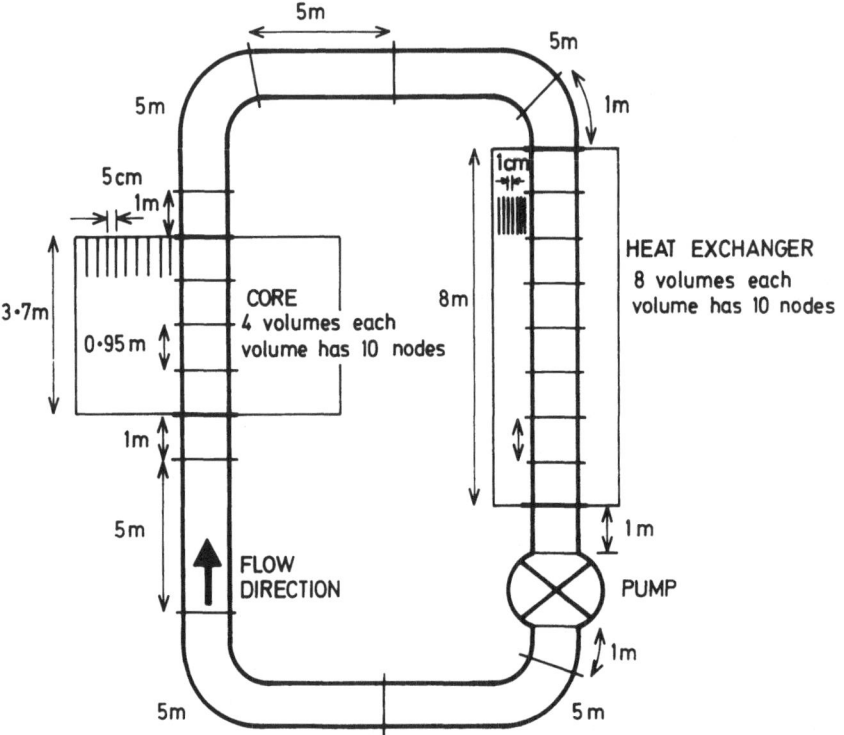

Closed circuit pipe is assumed to be of constant area.
Diameter of pipe = 10 cm
Velocity of flow at steady state = 4.8 m s^{-1}.

Figure 4

Simplified Reactor Model

Physically the core is represented by an annulus, the inner surface being bounded by the primary fluid (coolant), the outer by an insulator.

The form of heat transfer between core and fluid is forced convection, the heat transfer coefficient h being given by the Dittus-Boetler equation:

$$N_u = \frac{hD}{\lambda} = 0.023 \ P_r^{0.4} R_e^{0.8} \ .$$

Heat Exchanger. The Heat Exchanger is similar to the core model except that there is no heat production. Secondary fluid flow is not considered, as external boundary condition specifies a temperature.

Further coupling between the core and the fluid is being developed. A neutron kinetics model is being incorporated and in addition, the core composition is to be layered so as to simulate more closely a fuel rod with cladding.

The Plant data and the notation used is given below.

TABLE I

PLANT DATA

Heat generation rate (constant)	66.9 kW
Length of fuel rod	3.8 m
Fuel rod linear heat rating, average	17.8 kW/m
Core : Stainless steel (no cladding, no gaps)	
Coolant Pressure (abs)	172.4 bar
Coolant Temperature	350 °C
Coolant Pipe diamter	10 cm
Total Circuit Length	50.8 m
Friction Factor (constant)	0.02
Pump : Linear Characteristic	
Pump speed	1485 rev/min.
Heat exchanger height	8 m
Secondary coolant temperature (constant)	280 °C

Notation

c_v -   Heat capacity at constant volume

c_p -   Heat capacity at constant pressure

c   -   Specific heat capacity of heat conduction material

D   -   Diameter of pipe

h   -   Heat transfer coefficient

P   -   Pressure

T   -   Temperature

λ -   Heat conduction coefficient

u   -   Specific internal energy

t   -   Time

V   -   Velocity of fluid

α -   Ratio of time step to mesh length

ρ -   Density

f   -   Friction factor

β -   Isobaric coefficient of thermal expansion

κ -   Isothermal coefficient of thermal expansion

The use of parallel processing is discussed here only for the
solution of the hydrodynamic model, the details of which are
given below.

Model Details : Hydrodynamic Model

Figure 5 represents the control volumes (cells) denoted
by L, M, N etc., the junctions between the cells are denoted
by the numbers 1, 2, 3, etc. Assume junction 7 represents
the pump in that the velocity is specified at 7. It should
be noted that Figure 5 in fact depicts a closed loop since
the beginning and the end junctions are the same.

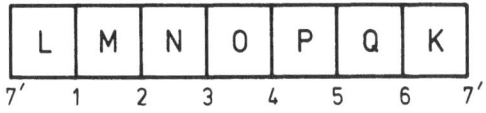

Figure 5

Control Volumes and Junctions

As in Section III, the equations are derived from considerations of continuity, energy and momentum. These equations are:

Continuity Equation:

$$\frac{\partial \rho}{\partial t} + \frac{\partial (\rho V)}{\partial x} = 0 \qquad (23)$$

Thermal Energy Equation:

$$\frac{\partial (\rho u)}{\partial t} + \frac{\partial (\rho V u)}{\partial x} = - P \frac{\partial V}{\partial x} + Q + \rho V^2 F_w \qquad (24)$$

Momentum Equation:

$$\rho \frac{dV}{dt} = \rho \frac{\partial V}{\partial t} + \frac{1}{2} \rho \frac{\partial}{\partial x} (V^2) = - \frac{\partial P}{\partial x} + \rho B_x - \rho V F_w,$$

where

B_x — being defined as the body acceleration in the x-direction i.e. a gravity term - $g\sin\theta$.

F_w — defined by $F_w = \frac{1}{2} \frac{f}{D} |V|$. For non-circular pipes, D is the equivalent hydraulic diameter, area/perimeter.

Q — defined as the heat input per unit volume.

f — friction factor depends on whether the flow is turbulent or laminar or in transition. We assume laminar flow.

Discretising the above equations in a similar fashion to RELAPV gives the following three equations per cell, where suffixes indicate time level whereas subfixes indicate cell volume letter or junction number.

Continuity:

$$\rho_M^{n+1} - \rho_N^n + \alpha_M \tilde{\rho}_M^n (V_2^{n+1} - V_1^{n+1}) = 0 \qquad (26)$$

Energy:

$$(\rho_M u_M)^{n+1} - (\rho_M u_M)^n + \alpha_M (\widetilde{\rho u})_M^n [V_2^{n+1} - V_1^{n+1}]$$

$$= - P_M^n [V_2^n - V_1^n]\alpha_M + Q_M^n \Delta t + \tfrac{1}{2}\tfrac{f}{D} \rho_M^n (V_M^n)^2 |V_M^n| \Delta_t$$

$$(27)$$

Momentum:

$$V_2^{n+1}[\rho_2^n + \tfrac{1}{2}\tfrac{f}{D} \rho_2^n |V_2^n| \Delta t] = \rho_2^n V_2^n - \tfrac{1}{2}\alpha_2 \rho_2^n [(V^2)_N^n -$$

$$(V^2)_M^n] + \tfrac{1}{2}\alpha_2 \rho_2^n \text{VISC}_2^n + \rho_2^n B_X \Delta t - \alpha_2 (P_N - P_M)^{n+1} \qquad (28)$$

where $\alpha_M = \Delta t / \Delta X_M$

$\alpha_2 = \Delta t / \Delta X_2$

$\text{VISC}_2^n = \tfrac{1}{2}[|V_N^n| (V_3^n - V_2^n) - |V_M^n| (V_2^n - V_1^n)]$,

$V_N = \tfrac{1}{2}(V_3 + V_2)$

All tilded quantities are donored quantities introduced to make the equations well posed:

$$\tilde{X}_M \triangleq \tfrac{1}{2}(X_M + X_L) + \tfrac{1}{2}\left| \frac{V_1}{V_1} \right| (X_L - X_M), \ V \neq 0$$

The integration scheme used in RELAPV has evolved from the need to achieve a fast execution speed. Implicit evaluation is used only for terms necessary for numerical

stability. Thus implicit evaluation is used for the velocity
in mass and energy transport terms and the pressure gradient
in the momentum equations. The need for linearity (which
results in high computing speed by not having to iteratively
solve systems of non-linear equations) required selecting
time-level evaluations so that the resulting implicit terms
become linear in the new time variables.

In addition to the above three equations, a fourth
equation (state relationship) is needed to make the system
determinant; a generalisation of the compressibility factor
of the previous example. In general,

$$\rho = \rho(P,u),$$

where ρ is usually a non-linear function. To preserve the
linearity of the numerical scheme, the density relationship
is expressed by a two-term Taylor series expansion in terms
of the other state variables about the old time level. This
gives:

$$\rho^{n+1} = \rho^n + (\frac{\partial\rho}{\partial P})^n_u (P^{n+1} - P^n) + (\frac{\partial\rho}{\partial u})^n_P (u^{n+1} - u^n)$$

To obtain a more manageable form, the energy variable is
expressed as the product (ρu):

$$\rho^{n+1} = \rho^n + (\frac{\partial\rho}{\partial P})^n_{\rho u} (P^{n+1} - P^n) + (\frac{\partial\rho}{\partial(\rho u)})^n_\rho (\rho u)^{n+1} - (\rho u)^n)$$

$$(29)$$

where

$$(\frac{\partial\rho}{\partial P})^n_{\rho u} = \rho^n (\frac{\partial\rho}{\partial P})^n_u / [\rho^n + u^n (\frac{\partial\rho}{\partial u})^n_P]$$

$$(\frac{\partial\rho}{\partial(\rho u)})^n_P = (\frac{\partial\rho}{\partial u})^n_P / [\rho^n + u^n (\frac{\partial\rho}{\partial u})^n_P]$$

From standard thermodynamic relations,

$$(\frac{\partial\bar{v}}{\partial P})_u = - \frac{\overline{v}\kappa c_v}{(c_P - P\bar{v}\beta)} \quad \text{where } \bar{v} = \frac{1}{\rho}$$

and

$$(\frac{\partial\bar{v}}{\partial u})_P = \frac{\bar{v}\beta}{(c_P - P\bar{v}\beta)}$$

where

$$c_v - c_p = -\frac{\beta^2 T \bar{v}}{\kappa} \qquad (30)$$

The isobaric coefficient of thermal expansion is

$$\beta = \frac{1}{\bar{v}}(\frac{\partial \bar{v}}{\partial T})_P$$

and the isothermal coefficient of compressibility

$$\kappa = -\frac{1}{\bar{v}}(\frac{\partial \bar{v}}{\partial T})_T.$$

As the fluid being considered will always be in the liquid
state, the variation in β and κ is small and so can be
neglected over the particular transients chosen. Similarly,
c_p is also considered constant and derived from the
enthalpy change

$$(h_2 - h_1) = c_p(T_2 - T_1) \quad \text{(at constant pressure)}.$$

The above coefficients are obtained from standard steam
tables for super cooled water, the values being averaged
around a pressure of 12.5 MPa and a temperature of 443 K.
c_v is obtained from eqn. (30). As the fluid is approximately
incompressible, the internal energy becomes a function of
temperature only, i.e.

$$du = c_v dT$$

Solution of Equations

Since the equations are linear, it is possible to
eliminate variables by substitution.

v^{n+1} is eliminated from eqns. (26) and (27) using
eqn. (28).

ρ^{n+1} is eliminated from eqn. (26) using eqn. (29), the
resulting equation being used to eliminate

$(\rho u)^{n+1}$ from the energy equation. The resulting
equations are given below for mesh M (see Figure 2).

On substituting eqn. (29) and eqn. (28) into the continuity equation gives:

$$\left[\left(\frac{\partial \rho}{\partial(\rho u)}\right)_P\right]_M^n (\rho u)_M^{n+1} = \left[\left(\frac{\partial \rho}{\partial(\rho u)}\right)_P\right]_M^n (\rho u)_M^n + \left[\left(\frac{\partial \rho}{\partial P}\right)_{\rho u}\right]_M^n P_M^n$$

$$-\alpha_M^n \rho_M^n \left[\frac{\gamma_2^n}{\beta_2^n} - \frac{\gamma_1^n}{\beta_1^n}\right] + \left\{\alpha_M^n \tilde{\rho}_M^n - \left(\frac{\alpha_1^n}{\beta_1^n} - \frac{\alpha_2^n}{\beta_2^n}\right) - \left[\left(\frac{\partial \rho}{\partial P}\right)_{\rho u}\right]_M^n\right\} P_M^{n+1}$$

$$+ \frac{\alpha_M \tilde{\rho}_M \alpha_1}{\beta_1^n} P_L^{n+1} + \frac{\alpha_M \tilde{\rho}_M^n \alpha_2}{\beta_2^n} P_N^{n+1} \qquad (31)$$

Substituting eqn. (28) into eqn. (27) gives:

$$(\rho u)_M^{n+1} = (\rho u)_M^n - \alpha_M (\tilde{\rho u})_M^n \left[\frac{\gamma_2}{\beta_1} - \frac{\gamma_1}{\beta_1}\right] + \mu_M^n - \alpha_M (\tilde{\rho u})_M^n \left[-\frac{\alpha_2}{\beta_2}\right] P_N^{n+1}$$

$$-\alpha_M (\tilde{\rho u})_M^n \left[\frac{\alpha_2}{\beta_2} + \frac{\alpha_1}{\beta_1}\right] P_M^{n+1} - \alpha_M (\tilde{\rho u})_M^n \left[\frac{\alpha_1}{\beta_1}\right] P_L^{n+1}$$

Substituting for $(\rho u)^{n+1}$ from the above equation into eqn. (31) gives

$$\left[-\frac{\alpha_M \tilde{\rho}_M^n[\alpha_1/\beta_1 + \alpha_2/\beta_2]}{\left[\left(\frac{\partial \rho}{\partial(\rho u)}\right)_P\right]_M^n} + \left[\left(\frac{\partial \rho}{\partial P}\right)_{\rho u}\right]_M^n + \alpha_M (\tilde{\rho u})_M^n \left(\frac{\alpha_2}{\beta_2} + \frac{\alpha_1}{\beta_1}\right)\right] P_M^{n+1}$$

$$+ \left[\frac{\alpha_M \tilde{\rho}_M \alpha_1}{\beta_1 \left[\left(\frac{\partial \rho}{\partial(\rho u)}\right)_P\right]_M^n} - \frac{\alpha_M (\tilde{\rho u})_M^n \alpha_1}{\beta_1^n}\right] P_L^{n+1} + \left[\frac{\alpha_M \tilde{\rho}_M^n \alpha_2}{\beta_2 \left[\frac{\partial \rho}{\partial(\rho u)_P}\right]_M^n} - \right.$$

$$\left.\frac{\alpha_M (\tilde{\rho u})_M^n \alpha_2}{\beta_2}\right] P_N^{n+1} =$$

$$= \rho_M u_M^n - \alpha_M (\tilde{\rho u})_M^n \left[\frac{\gamma_2^n}{\beta_2^n} - \frac{\gamma_1^n}{\beta_1^n} \right] + \mu_M^n - TOR_M^n + \frac{\alpha_M \rho_M^n \gamma_2^n}{\beta_2 \left[(\frac{\partial \rho}{\partial (\rho u)})_P \right]_M^n}$$

where

$$TOR_M^n = \frac{\left[(\frac{\partial \rho}{\partial (\rho u)})_P \right]_M^n (\rho u)_M^n + \left[(\frac{\partial \rho}{\partial P})_{\rho u} \right]_M^n P_M^n - \alpha_M \tilde{\rho}_M^n \left[- \frac{\gamma_1^n}{\beta_1^n} \right]}{\left[(\frac{\partial \rho}{\partial (\rho u)})_P \right]_M^n}$$

$$\left[(\frac{\partial \rho}{\partial u})_P \right]_M^n = - \frac{\rho_M^n \beta}{[c_P - P_M^n \beta / \rho_M^n]} \quad ; \quad \left[(\frac{\partial \rho}{\partial P})_u \right]_M^n = \frac{\rho_M^n \kappa c_v}{[c_P - P_M^n \beta / \rho_M^n]}$$

$$\beta_1^n = \rho_1^n \left[1 + \tfrac{1}{2} \frac{f}{D} |v_1^n| \Delta t \right] \quad ; \quad \beta_2 = \rho_2^n \left[1 + \tfrac{1}{2} \frac{f}{D} |v_2^n| \Delta t \right]$$

$$\gamma_1^n = \rho_1^n v_1^n - \tfrac{1}{2}\alpha_1 \rho_1^n \left[(v^2)_M^n - (v^2)_L^n \right] + \tfrac{1}{2}\alpha_1 \rho_1^n VISC_1^n + \rho_1^n B_x^n \Delta t$$

$$\gamma_2^n = \rho_2^n v_2^n - \tfrac{1}{2}\alpha_2 \rho_2^n \left[(v^2)_N^n - (v^2)_M^n \right] + \tfrac{1}{2}\alpha_2 \rho_2^n VISC_2^n + \rho_2^n B_x^n \Delta t$$

$$\mu_M^n = -P_M^n [v_2^n - v_1^n] \alpha_M + Q_n^n \Delta t + \tfrac{1}{2}\rho_M V_M^2 \frac{f}{D} |v_m| \Delta t$$

$$\tilde{\rho}_M^n = \tfrac{1}{2} (\rho_M + \rho_L) + \tfrac{1}{2} \left| \frac{V_1}{V_1} \right| (\rho_L - \rho_M)$$

$$v_M^n = \tfrac{1}{2} (v_1^n + v_2^n)$$

$$VISC_1^n = \tfrac{1}{2} \left[|v_M^n| \ (v_2^n - v_1^n) + |v_L^n| \ (v_1 - v_7) \right] ,$$

$$VISC_2^n = \tfrac{1}{2} \left[|v_N^n| \ (v_3^n - v_2^n) + |v_M^n| \ (v_2^n - v_1^n) \right] .$$

V_7 and P_L^{n+1} are boundary conditions specified at the pump.

Similarly, for mesh K the resulting equation after substitution is

$$\left[-\frac{\alpha_x \rho_x^{\sim n} \alpha_6/\beta_6 + [(\frac{\partial \rho}{\partial P})_{\rho u}]_K^n}{[(\frac{\partial \rho}{\partial (\rho u)})_P]_K^n} + \frac{\alpha_x (\widetilde{\rho u})_K^n \alpha_6}{\beta_6^n} \right] P_K^{n+1} +$$

$$\left[\frac{\alpha_K \rho_K^{\sim n} \alpha_6}{\beta_6 \ [(\frac{\partial \rho}{\partial (\rho u)})_P]_K^n} - \frac{\alpha_K (\widetilde{\rho u})_K^n \alpha_6}{\beta_6^n} \right] P_Q^{n+1}$$

$$= \rho_K^n u_K^n - \alpha_K (\widetilde{\rho u})_K^n \left[v_7^{n+1} - \frac{\gamma_6^n}{\beta_6^n} \right] + \mu_K^n - TOR_K^n + \frac{\alpha_K \rho_K v_7^{n+1}}{[(\frac{\partial \rho}{\partial (\rho u)})_P]_K^n}$$

where

$$TOR_K^n = \left[\frac{(\frac{\partial \rho}{\partial (\rho u) P})_K^n (\rho u)_K^n + [(\frac{\partial \rho}{\partial P})_{\rho u}]_K^n P_K^n - \alpha_K \rho_K^{\sim n} - \gamma_6^n/\beta_6^n}{[(\frac{\partial \rho}{\partial (\rho u)})_P]_K^n} \right]$$

$$VISC_6^n = \tfrac{1}{2} \left[|v_K^n| \ (v_7 - v_6) + |v_Q^n| \ (v_6 - v_5) \right] .$$

V_7 is the boundary condition specified by the pump.

Thus the solution can be advanced in time by solving the following form of matrix equation in terms of pressure:

$$
\begin{bmatrix}
X & X & & & & & 0 \\
X & X & X & & & & \\
& X & X & X & & & \\
& & X & X & X & & \\
& & & X & X & X & \\
0 & & & & X & X \\
\end{bmatrix}
\begin{bmatrix}
P_M^{n+1} \\
P_N^{n+1} \\
P_O^{n+1} \\
P_P^{n+1} \\
P_Q^{n+1} \\
P_K^{n+1}
\end{bmatrix}
=
\begin{bmatrix}
X \\
X \\
X \\
X \\
X \\
X
\end{bmatrix}
\qquad (32)
$$

where the Xs represent non-zero elements. Note that the matrix is tri-diagonal. Once the above matrix equation has been solved in terms of P^{n+1} all other scalar and vector properties can be obtained by back substitution.

The time step control of the simplified system is based upon the error measure

$$
\varepsilon = \frac{1}{N} \sum_{i=1}^{N} \left| (\rho_i - \rho_{mi})/\rho_i \right|
$$

where ρ_i is the density calculated from the state equation and ρ_{mi} is the density calculated from the continuity equation. If the above error falls below the acceptable error band, the time step can be doubled while if it is greater than the acceptable error band the time step is reduced by a factor of two and the calculation is repeated. The heat transfer model is also linear in advanced time and is similar to eqn. (32).

In the model no time step control exists for the heat transfer calculation as the fluid system is assumed to be a more stiff system. In addition, if reactor kinetics had been included, these would also have controlled the advancement of the heat transfer system.

The solution of eqn. (32) forms only a part of the computational load when advancing one time step. Here, we consider how this equation may be solved using parallel processing. Eqn. (32) is of the form

$$AP = b \qquad (33)$$

where A is a tri-diagonal matrix. The above equation can be partitioned into ν sets of equations of the form:

$$A_{ii}P_i + y_i = b_i$$

$$y_i = A_{i,i-1}P_{i-1} + A_{i,i-1}P_{i+1}$$

$$i=1,2,\ldots.\nu \qquad (34)$$

where $A = \begin{pmatrix} A_{11} & A_{12}\ldots.A_{1\nu} \\ A_{21} & A_{22}\ldots.A_{2\nu} \\ A_{\nu 1} & A_{\nu 2}\ldots.A_{\nu\nu} \end{pmatrix}$, $P = \begin{pmatrix} P_1 \\ P_{.2} \\ \cdot \\ P_\nu \end{pmatrix}$ and $b = \begin{pmatrix} b_1 \\ b_{.2} \\ \cdot \\ b_\nu \end{pmatrix}$.

The partitioned eqn. (34) is similar to the one given in Section II. It is thus possible to use the decomposition coordination technique given in that Section. The coordinator would supply values of the variables y_i, i=1,2,....ν, and given these values the variables P_i, i=1,2.....ν may be computed in parallel (on ν processors). The coordinator can use either of the two strategies (relaxation or Newton-Raphson) discussed in Section II. The advantage of using Newton-Raphson, however, is that only one iteration would be required. (See comments at the end of Section II).

Detailed simulation studies are currently being carried out to assess the above approach. Initial results obtained when the Newton-Raphson procedure is used in the coordinator are as given below:

No decomposition   :   21 time units
With decomposition, ν = 2 :   24 time units
With decomposition, ν = 3 :   27 time units
With decomposition, ν = 4 :   29 time units

The times shown above are for a serial machine, so that if ν parallel processors were used, then the times would be ν times less than those given above.

V. CONCLUSIONS

In this Chapter, the problem of rapid simulation as
required for example in the nuclear industry has been
considered. The approach has been to make use of a decomposition-coordination technique which enables the overall
problem to be split into parallel tasks, so that parallel
computing facilities can be used.

The method discussed has been applied to two examples.
These show the potential of the decomposition-coordination
techniques for solving equations resulting from a set of
PDEs or ODEs. The second example considered in this chapter
is particularly relevant to the RELAPV nuclear code. However,
more work remains to be done before an accurate assessment
of the possibilities of speeding up this code using decomposition-coordination techniques can be made. It is necessary
to include in the reactor model being considered some of the
features that have been neglected initially. Present work
is concerned with this issue.

ACKNOWLEDGEMENTS

The authors would like to thank Dr K Malinowski for
collaboration in the general area of system simulation, and
Mr R. Buro and Mr W. Crorkin for the simulation examples
given in this paper.

REFERENCES

[1] Allidina A.Y. (ed.): 'Development of hierarchical
     techniques for the simulation of large scale systems
     with particular application to the nuclear industry',
     EEC Project Phase 1 Report, May, 1984.

[2] Allidina A.Y., Lei S. and Wang L.: "Hierarchical
     simulation techniques for ODE Systems', Distributed
     Simulation 85, San Diego, California, Jan. 1985.

[3] Arnold C.P., Michael I.P. and Michael B.D.: 'An
     efficient parallel algorithm for the solution of large
     sparse linear matrix equations'. IEEE Trans. on
     Computers, Vol. C.32, No. 3 (March 1983).

[4] Barlow R.H. and Evans D.J.: 'Parallel algorithms for
     the iterative solution of Linear Systems', The Computer
     Journal, Vol. 25, No. 1, 1982.

[5] Brasch F.M. Jr., Van Ness J.E. and Kang S.C.: 'Design
     of Multiprocessors structures for simulation of power-
     system dynamics', Report, Electric Power Research
     Institute, Palo Alto, California 94304, U.S.A.

[6] Burks A.W.: 'Programming and structure changes in
     parallel computers', Proceedings CONPAR 81, Springer-
     Verlag, Lecture Notes in Computer Science, Vol. III,
     edited by W. Handler, 1981.

[7] Evans D.J. and Haghighi R.S.: 'Parallel iterative
     methods for solving linear equations'. Intern. J.
     Computer Math., Vol. II, pp 247-285, 1982.

[8] Findeisen W., Bailey F.N., Bryds M., Malinowski K.,
     Tatjewski P. and Wozniak A.: 'Control and coordination
     in hierarchical systems'. International Series on
     Applied Systems Analysis, Vol. 9, John Wiley & Sons, 1980.

[9] Franklin, M.A.: 'Parallel solution of ordinary
     differential equations', IEEE Trans. on Computers,
     Vol. C-27, No. 5, May 1978.

[10] Halada L.: 'A parallel algorithm for solving band
     systems and matrix inversion', Proc. CONPAR 81, Springer-
     Verlag, Lecture Notes in Computer Science, Vol. III,
     edited by W. Handler, 1981.

[11] Himmelblau D.M. (editor): 'Decomposition of large scale
     systems'. Collection of articles on decomposition and
     coordination techniques). American Elsevier, New York
     1973.

[12] Katz. I.N., Franklin M.A. and Sen A.: 'Optimally stable
     parallel predictors for Adams-Moulton correctors',
     Comp. and Maths with Appls., Vol 3, pp. 217-233, 1977.

[13] Lasdon L.S.: 'Optimisation Theory for Large Systems',
     MacMillan, London, 1970.

[14] Malinowski K., Allidina A.Y., Singh M.G. and Crorkin W.:
     "Decomposition-coordination techniques for parallel

simulation - Part 1', Control Systems Centre Report No. 599, UMIST, Manchester, March 1984.

[15] Miranker W.L. and Liniger W.: 'Parallel methods for the numerical integration of ordinary differential equations', Math. Comput. Vol. 21, pp 303-320, 1967.

[16] RELAPV Code Manuals Vols. 1,2 and 3, U.S. Department of Energy, 1982.

[17] Schendel U.: 'On basic concepts in parallel numerical mathematics', Proc. CONPAR 81, Springer-Verlag, Lecture Notes in Computer Science, Vol. III, edited by W. Handler, 1981.

[18] Singh M.G. and Titli A.: 'Systems: Decomposition, Optimisation and Control', Pergamon Press, Oxford 1978.

[19] Singh M.G., Allidina A.Y. and Malinowski K.: 'Hierarchical simulation techniques', MECO 83, Athens, July, 1983.

[20] Travassos R. and Kaufman H.: 'Parallel algorithms for solving non-linear two-point boundary-value problems which arise in optimal control', Journal of Optimisation Theory and Applications, Vol. 30, No. 1, January, 1980.

[21] Wismer D.A.: 'Distributed multilevel systems', in Optimisation Methods for Large-Scale Systems, edited by D.A. Wismer, McGraw-Hill, New York, 1971.

[22] Worland P.B.: 'Parallel methods for the numerical solution of ordinary differential equations', IEEE Trans. on Computers, October, 1976.

DEVELOPMENTS IN FULL-SCOPE, REAL-TIME NUCLEAR PLANT SIMULATORS

J. Wiltshire

Marconi Instruments
Donibristle
Scotland

I. INTRODUCTION

Simulation techniques have been used in the nuclear industry since its inception for design and safety study purposes. Many of the early simulators were analogue computers which could provide high processing speeds but were limited in capacity and repeatability. These systems gave way to digital/analogue hybrids which overcame some of the limitations and these in turn were superceded by entirely digital systems.

Digital computers generally have passed rapidly through a series of generations and a number of powerful general purpose simulation packages are in constant use on both mainframes and supermini-computers. A general trend with these systems has been towards preservation of the underlying computing environment and to rely upon a periodic hardware upgrade to achieve a performance compatible with the requirement.

Developments in device technology and machine architecture have produced a dramatic rate of improvement which acts in support of this policy and has masked the need for an alternative approach. There is however, a fundamental limit to the speed at which the information pipeline implicit in a mainframe can operate and modern systems are approaching this limit.

It may be supposed that the issues involved are essentially concerned with microelectronics, computer design

191

principles, language structures and similar, related topics.
This is true to a degree but the Nuclear Plant simulation
activity should be viewed as a "total process" if it is to be
properly understood and this is a formidable task considering
the range of different disciplines involved. However, a
number of apparently "obvious" observations can be made.

If computing capacity becomes a problem then why not
operate two (or more) computers in parallel?

Initially the answer to this question was purely a
matter of cost but years of development in the single mainframe mold has evolved mathematical and organisational
techniques for which parallelism was never a consideration.
Consequently, the technique is unpopular and discussions on
laws of diminishing returns due to communications overheads
and predictions of unreliability are common.

The advent of the microprocessor has dramatically reduced
the cost of multiprocessor systems and this, in part, has
provided the motivation for a challenge to this viewpoint.
There is, however, a more powerful source of motivation for
research into the use of parallel computation in Nuclear
Simulation.

Digital simulations used for design or safety study
purposes are not invalidated if the computer runs slowly or
even disjointly as in a multiprogramming system. Many big
models are run overnight or at weekends. However, simulators
used for training purposes must produce continuous results
interactively, without prior knowledge of operator actions
and at the same rate as the real plant. They must also have
a dynamic range capable of supporting training exercises as
continuous processes and the required operating envelopes
are often much wider than that appropriate to the "set piece"
approach common in design simulation.

It has been argued that training simulators do not require
the in-depth physical representations built into the reference
codes and there are areas where this is indeed the case.
However, simplified models which are not obviously wrong are
very difficult to design and they introduce uncertainties into
the validity of the training offered.

Conservation of a Quality Assurance route from the design
models into the training simulator is an extremely attractive

objective but it requires real time implementation of the
design models themselves.

In consequence the design of a large, Nuclear Plant
training simulator must make provision for a formidable computing load.

In making the transition from the simulation of small
Pressurised Water Reactors (PWR) to the Advanced Gas Cooled
Reactor (AGR) civil power plants, Marconi Instruments has
developed a system design philosophy appropriate to computer
configurations which exhibit a high degree of parallelism.
The Hunterston B AGR simulator entered service in 1984 and a
further system is currently under construction for the new
Torness plant.

These systems use large numbers of microprocessors
(Hunterston 52, Torness 48) operating in parallel to evaluate
extensive mathematical models of the respective plants.

The systems design philosophy represents a departure from
previous practice and is particularly wide ranging. The simulators are seen as systems in which engineering disciplines
drawn from widely different fields come together and interact
in a complex fashion. Optimisation of these interactions is
seen as a major priority and this results in the identification of "ground rules" peculiar to real time simulator design.

II. NUCLEAR SUBMARINE SIMULATORS

In 1964 the Royal Navy identified the need for a simulator for use in part of the training programme for the
engineering watchkeepers in their Nuclear Submarine service.
Their first system was delivered in 1967 and it consisted of
a digital computer interfaced to a replica machinery control
room. The computer evaluates a mathematical model of the
reactor, the electrical system and the propulsion system.
This evaluation takes place at real time and it interacts
with the operator controls and indications via extensive
interface equipment.

To date, a total of seven such simulators have been
delivered and the Royal Navy has accumulated considerable
operational experience. All of these systems are designed to
re-create the environment of a particular class of boat and
in this respect they are plant specific. Some generic

training is carried out to augment classroom lessons but the main function of the simulators is in the full-scope mode. This expression relates to a mode of operation in which all relevant component subsystems of the plant are represented and they interact to provide an integrated environment of maximum realism. Audio simulation of diesels, circuit breakers and steam leaks provide action related cues which further enhance the level of simulation. With an experienced instructor manning the control console and intercom, engineering watchkeepers accept these machines "as real" in their response to both normal and emergency situations.

III.   THE HUNTERSTON 'B' SIMULATOR

Hunterston 'B' is an Advanced Gas-cooled Reactor Nuclear Power station situated on the Firth of Clyde in Scotland and operated by the South of Scotland Electricity Board (SSEB). A total generating capacity of 1320 MWe is distributed between two reactor-turbogenerator units which entered service in 1976 and 1977. Following normal practice, the SSEB recruited the operations staff at an early stage in order to employ them in the dual role of future plant operators and commissioning engineers. As a result, operators obtained in-depth knowledge of the plant and its behaviour under a variety of conditions.

After some years of base-load operation, SSEB felt that it was important to give operators the opportunity to rehearse other, less common modes of operation and became interested in the concept of a high fidelity training simulator to supplement the generic and theoretical training received at the National Nuclear Power training centre operated by the Central Electricity Generating Board (CEGB). A number of possible locations were considered for the Hunterston 'B' simulator including the SSEB's training centre and it was concluded that the balance of advantage lay with locating the facility at site because of staff availability and the capacity to carry out site emergency training without disturbance to the running plant.

After a series of design studies, SSEB placed a contract in July 1981 with Marconi Instruments, Ltd. to provide a site-based full scope simulator for one reactor-turbine unit.

The simulated control room, computer room and offices are being contained within a specially constructed building separate from the main plant but within the nuclear site licensed boundary.

A. Functional Specification

At Hunterston 'B' all first-line control actions are carried out from the central control room and the operator interface is substantially conventional.

The fundamental specification of the simulator is that from the perspective of an operator in its control rooms, the simulator should look like, behave like and indeed "feel" like the real plant.

The specification includes an extensive set of training exercises which cover both normal and emergency operation of the plant. Some of these are listed in Table I.

This list results in a clear definition of the extent and scope of the simulation required. For example, the requirement for the simulator to respond sensibly to bulk gas inlet temperature and mass flow perturbations, in a total plant sense, requires that a reasonably accurate axial representation of the neutron kinetic and thermodynamic relationships be established. In the same way, the require-ment that the simulator should respond adequately to a boiler quadrant trip and reinstatement; reactor trimming; group; sector and single rod withdrawal faults; single channel flow faults and refuelling establishes the requirement for explicit modelling of individual fuel channels and boiler units.

The scope of balance of plant simulation includes exten-sive representation of the turbine-generator with its auxiliaries such as gland steam, lubricating oil, condenser, vacuum plant and condensate heat recovery systems, together with the governor and automatic voltage regulator.

A detailed simulation of all four direct contact feed heaters with their pumps and level control system combine with the feed pump models, comprising turbine and electric motor driven back-up cooling system pumps, to complete the thermodynamic cycle.

Electrical auxiliary supplies are included in the form of the 23 kV, 11 kV, 3.3 kV and 415 kV systems.

All inter-plant and operator-plant interfaces are inclu-ded in detail and these cover autocontrols, plant protection alarms, monitoring and post trip logic systems.

TABLE I

HUNTERSTON 'B' : SIMULATOR FUNCTIONAL SPECIFICATION

(1) Start-up/shut down.

(2) On-load fuelling.

(3) Temperature/power trimming.

(4) Turbine warm-up and run-up.

(5) Deaerator temperature control.

(6) Operation of the Reactor Shutdown Sequence Equipment.

(7) Reinstatement of a boiler quadrant.

(8) Loss of main boiler feed pump.

(9) Loss of one out of two standby boiler feed pumps.

(10) Loss of boiler quadrant.

(11) Loss of gas flow.

(12) Control rod withdrawal fault at start-up.

(13) Loss of control loops.

(14) Loss of feed heaters; partial.

(15) Loss of feed heaters; complete.

(16) Loss of single station transformer.

(17) Loss of reactor Cooling Water (CW) system.

(18) Loss of reactor auxiliary cooling water system.

(19) Blocked channel incidents.

(20) Boiler leaks - small/large.

(21) Depressurization faults - with/without grid.

(22) Loss of main CW system.

(23) Total loss of station supplies.

(24) Total loss of normal feed system.

(25) Charge machine emergency cooling system faults.

(26) Gas baffle failure.

B. Design Approach

It was appreciated that the achievement of the stated requirements would be difficult unless the amount of empirical modelling was minimized by the use of models based upon fundamental physics and engineering. Similarly, all plant logic systems should be represented by accurate simulation of the plant drawings to ensure that the same logic applies. It was also decided to base the simulator, as far as possible, on existing design models which had undergone considerable rig and plant correlation and, in particular on those developed by the CEGB's Nuclear Plant Kinetics Group.

The decision was therefore made to implement existing design models, extended and augmented to meet real-time fully interactive total plant requirements.

The design approach not only conserves in the simulator the existing extensive design model validation experience, but it should also enable advantage to be taken of any future improvements in the design data base.

As far as possible, the simulator was commissioned using the same procedures that were used for commissioning the real plant and experienced Hunterston 'B' operators were used to comment on the fidelity of the responses produced.

In some areas where no validated design models existed, mathematical representations based on physical conservation laws were devised in a collaborative effort by the SSEB and Marconi Instruments with assistance from the CEGB, the most significant examples being the development of the asymmetric reactor, primary circuit and boiler representation.

C. Asymmetry

In the Hunterston-style Advanced Gas Reactor, the primary gas coolant and boiler feed are controlled on a quadrant basis although there are two gas circulators and six boiler half-units within each quadrant which are subject to manual trim and auto-control actions.

circulators, and three low resolution quadrants of the
reactor core in each of which the reactor is represented by
four fuel channels per mesh point and the boiler by a single
representative tube. This arrangement is shown in Figure 1.
Although the fluxes and powers are thus calculated at 136
mesh points, they are mapped back into the reactor geometry
to yield individual channel flows and outlet gas temperatures. Minor asymmetries generated by the mixed geometry
are acceptably small and the model compares favorably with
the responses of a point reactor model used for design
studies of symmetric transients.

In most circumstances, the axial flux shape in the
reactor tends to be radially and azimuthally invariant and
thus the two-dimensional model can be combined with the axial
representation shown in Figure 2 to produce a synthesized
three-dimensional model. This provides the simulator with
the capability to represent the effects of individual rod
and gag trimming and the asymmetry associated with a variety
of operational plant configurations.

Approximately once per month, station staff perform a
detailed snapshot assessment of the reactor core state for
the purposes of fuel records and temperature assessment. A
data route has been set up whereby such core state information
can be transferred to the simulator and the snapshot channel
powers, temperatures, flow and nuclear properties conserved.
This "core following" capability is also dependent upon the
simulator ability to represent the asymmetries inherent in
normal plant operation.

D. Instructor Facilities

The simulator is located in a specially designed building, the internal layout being exactly the same size and
shape as the plant control room with an identical disposition
of the desks and panels. To fulfill the requirement of
running site emergency exercises from this location, the
simulator control room is equipped with a full complement of
communication systems including public address, direct wire
and PABX telephones as well as radios.

Considerable attention has been given to the role of the instructor and the simulator design includes a number of facilities to assist in the generation, monitoring and control of complex training exercise scenarios.

Although the reactor core contains 308 fuel channels, the refuelling cycle is managed to main quadrantal symmetry as far as possible. It is possible, therefore, to represent the complete steam generating system by one high resolution quadrant of the reactor core in which individual fuel channels are modelled along with six boiler half-units and two gas-

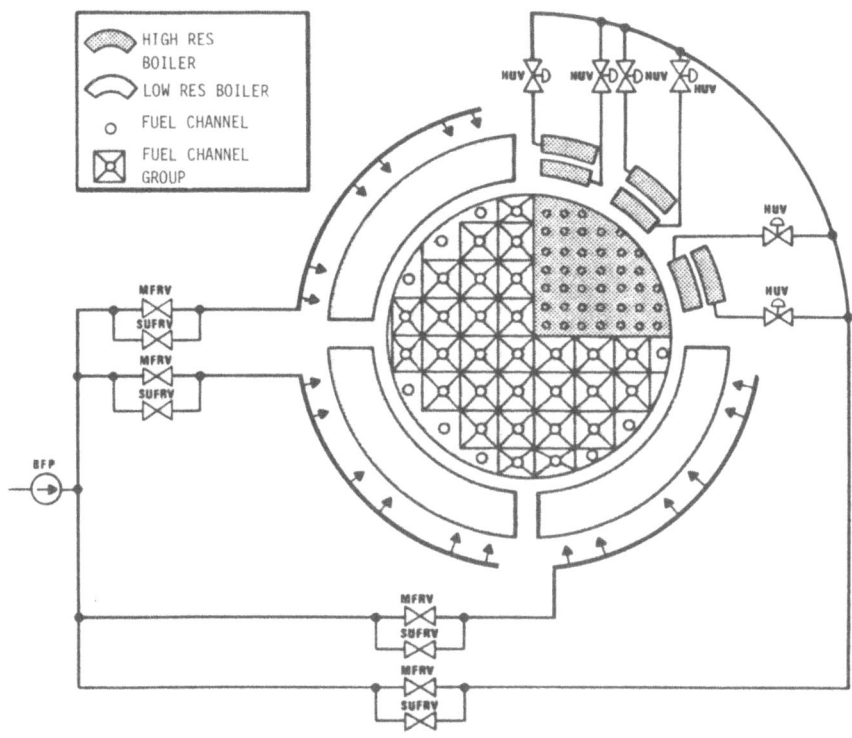

Figure 1.   Dual Resolution Asymmetric Reactor/Boiler Model

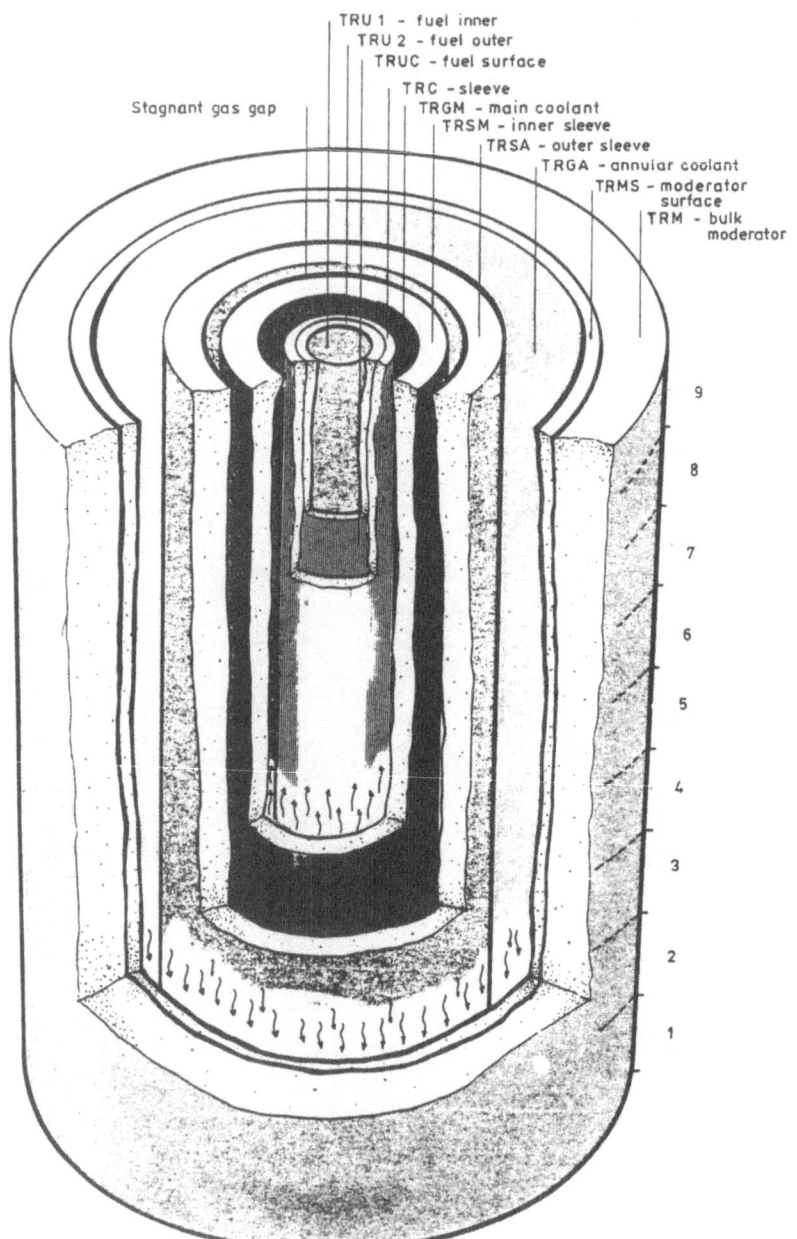

Figure 2.   Axial Reactor Model

The instructors' room is situated behind and above the dummy control panels of the second non-simulated unit and provides a clear view of the whole control room. This allows the instructor to control and observe exercises without intruding into the simulation.

The equipment used by the instructors is contained in two consoles which include their own local computer systems. The two consoles can be used independently for separate concurrent exercises, or in combination with two instructors interacting within the same exercise. The inclusion of a computer and high capacity disc system in each console supports a number of powerful control and monitoring facilities.

These facilities enable fault sequences or plant perturbations to be activated automatically or following cues given by the instructor. In addition, provision is made for direct control of faults and simulation control options through a generalized console based on DIN compatible modules which provide conventional digital and analogue controls. Association between these controls and the simulator's data base can be established to give direct and explicit control of faults, error functions and indications.

A series of color graphic mimics provide fully animated views of major parameters and these can be called up by the instructor for the rapid assessment of operator actions and consequential plant states. The mimics have also proved invaluable in developing and commissioning some of the models. For example, perspective projections of flux and temperature surfaces for the two-dimensional reactor model were invaluable in assessing the time and spacial convergency properties of these very large models.

In general, the instructor facilities were designed to be capable of supporting specific methods of monitoring and controlling complex training exercises without being committed to any particular approach. The underlying capabilities are, therefore, essentially open-ended and capable of being optimized to suit the instructor's particular and possibly changing requirements.

E. Implementation

The central issue in meeting the specification of the Hunterston 'B' simulator is the achievement of design quality

mathematical models, operating not only in real time but
also within an operating envelope which extends up to the
point at which the plant physics become uncertain, e.g.
fuel cladding meltout. While such excursions are normally
inhibited by the plant protection and safety systems, the
functional specification of the simulator called for exercises
in which these elements are deemed to have failed. Design
codes seldom include extensive representation of the logical
elements of a plant and they concern themselves mainly with
the analogue components. This results in a requirement to
solve large sets of differential, algebraic and Boolean
equations in real time. A summary of the distribution of
these equation types is shown in Table II.

A total of 52 microprocessors operating in parallel were
used in the Hunterston simulators to achieve real time operation. All of these machines are identical, interchangeable
and capable of independent operation. In general terms a
conventional single address architecture is employed in the
processor design which is microprogrammed to operate on
Boolean, integer and floating point variables. The architecture is designed to bear a one-to-one relationship with
the programming language used, making machine code facilities
unnecessary. However, in order to optimise certain processes
a number of generic microprograms are included in the processor design which accelerate commonly used algorithms.
Such processes include alarm processing, neutron diffusion,
the solution of differential equations, gas transport,
arbitrary function generation and interprocessor communication. The allocation of processors to elements of the asymmetric Reactor model is shown in Figure 3.

During the course of the project a series of experiments
were conducted using the computer system both in independent
sections and as an integrated whole. These experiments
established a methodology which included hardware, software
and mathematical techniques appropriate to the real time
solution of large models by systems which exhibit a high
degree of parallelism. These techniques were applied in
specific terms to Hunterston and then generalised for use on
the Torness simulator.

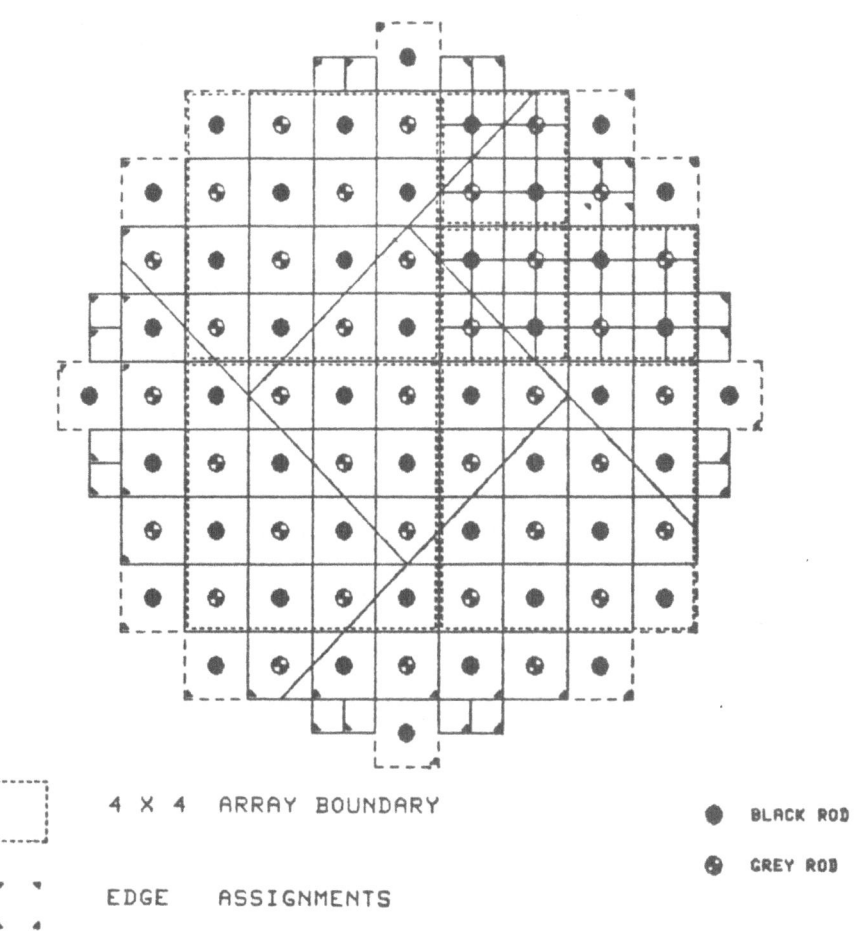

Figure 3. Asymmetric Reactor Model Processor Allocation

TABLE II

HUNTERSTON/TORNESS EQUATION RELATIONSHIPS

	Hunterston 'B'			Torness		
	Diff. Eqns.	Alg. Eqns.	Boolean Eqns.	Diff. Eqns.	Alg. Eqns.	Boolean Eqns.
Reactor (x, y)	3359	2720	–	7636	6640	–
Reactor Axial	353	180	–	353	180	–
Rod controllers	20	730	–	225	765	–
Gag flow	–	308	–	–	332	–
Gas circulators	20	308	–	20	333	–
Circulators	24	96	–	24	96	–
Electrical system	20	1000	–	20	1000	–
Boilers	1170	2700	–	1690	3900	–
Turbine	120	550	–	120	550	–
Control systems	33	200	–	33	200	–
Protection system	–	–	5000	–	–	5000
Feed pumps	10	50	–	10	50	–
Gas clean up	–	–	–	20	50	–
TOTALS	5129	8842	5000	10151	14096	5000

IV. THE TORNESS AGR SIMULATOR

Torness is the SSEBs second AGR plant and it is situated on the east coast of Scotland. While the underlying design principles are generic to all AGRs, many detailed differences exist between the two stations. Torness also includes an extensive multi-computer system for plant monitoring and auto-control purposes. An identical set of computers are included in the Marconi Instruments simulator which has an engineering role in addition to its use in training plant operators.

As can be seen in Table II, the Torness design requires a significantly increased computing capacity. This is provided by a set of 48 upgraded Marconi computers which are upwards compatible with the Hunterston standard both in terms of code execution, peripheral devices and mechanical assembly.

Since the simulator design phase preceded power raising on the station, operational experience of the specific Torness plant was not available. However, the compatibility of the two computer systems allowed the Hunterston model implementations to be used as working prototypes in order to make a controlled transition to the Torness requirement.

This process reduces the amount of new code being tested in the full-scope context and provides closed-loop boundary conditions corresponding to total plant perturbations.

V. SIMULATOR SYSTEM DESIGN PHILOSOPHY

Experience in designing and setting to work the Hunterston and Torness simulators has led to the evolution of a system design philosophy appropriate to large real time systems. A high degree of parallelism is implicit in the mathematical formulations of the models and in the hardware required to evaluate them. A simulator project is seen as an integrated entity with many of the conventional boundaries between the component disciplines weakened or removed completely. An outlook is engendered which reflects the inherent characteristics of real time, plant specific simulation which is treated as a subject in its own right. This reduces the number of inter-discipline interfaces and allows the requirements of one aspect of the engineering task to be accurately matched with the otherwise hidden capabilities of another. The prime example of this is the construction of

computer systems with an architecture derived from the
mathematical properties of a large nuclear plant model and
yet retaining a high utility when used for general software
commissioning.

A. Model Classification

In order that a clear distinction can be drawn between
the different classes of mathematical model required for an
accurate representation of a real plant, two major classes
of model have been identified and are shown in the following
paragraphs.

Physics Based Models. Plant elements such as the Reactor,
Boilers and Turbine are essentially physics based since their
performance can only be inferred from the laws of physics and
the properties of the materials involved. On a smaller scale,
other physics-based elements can be discerned which occur in
great profusion throughout the plant. These include valves,
pumps, vessels and busbars. While the physics of these elements is relatively simple, the number of variations on a
particular theme are considerable. However, if the models
involved are developed as generic elements which can be called
up and conditioned for a specific local purpose, then their
number reduces dramatically.

The essential principles upon which such models are based
cannot usually be derived from a study of the plant construction information since this is concerned with manufacturing
structures rather than plant performance.

The model design approach therefore consists of deriving
physics based models in general terms from the conservation
laws and appropriate formulations of the physical properties
of the material. Detailed plant information in the form of
capacities, geometries and material types is then required
to configure a specific instance of the general formulation.

Since the same principles apply to the plant design for
which a number of suitable industry standard codes exist, such
codes constitute the design baseline rather than plant construction information.

Information Based Models. In many areas the converse of
the above situation is the case and the required model is fully
defined by the plant design information. The best example of

this is relay logic for which the plant circuit diagram is a complete specification of, the model, its test specification and its relationship to other elements of the simulation. Nothing more than the symbol convention and basic electrical characteristics of the relays and their contacts are required.

Movement away from pure logic and towards the physics-based elements progressively introduces the issues discussed in that section. However, if the associated physics elements can be taken as read, then many more plant areas become purely information based.

Since many of these physics elements are well established and apply many times throughout the plant the identification of a suitable generic representation is appropriate. Once these generics have been established large areas of the models become purely information based.

System design aspects in this area are, therefore, covered by the provision of a suitable set of generics.

VI. AUTOMATIC TESTING

A full-scope simulator must have the capability to operate as an integrated entity and in real time. The performance of the complete system is dependent upon that of its component parts in such a manner that fault diagnosis from the full scope viewpoint can be a very complex process.

It is necessary to establish a high level of confidence in all components at a number of levels. The base level should establish the fundamental characteristics of appropriately sized packages. A number of such packages can then be combined into a higher level package and so on until a complete hierarchy is established to form the simulator system. These packages are referred to as Testable Units (TUs). They usually consist of programs which evaluate the mathematical models of plant items and they have to be extensively tested as standalone entities. This testing process often involves the application of predefined forcing functions which induce transient responses from the model. These responses are then compared with reference data and the performance of the model is assessed accordingly.

In order that the common elements of this process can be provided by standard utility software the concept of an

Automatic Test Environment (ATE) is used to support the
development of test software.

While the details of specific tests must remain dedicated
to the particular model involved, the mechanisms for the control and recording of test transients can be standardised.
In consequence the Automatic Test Environments are dedicated
programs which CALL the utilities as sub-routines.

The data produced by running the model within its unique
ATE can then be compared with predefined reference data. The
comparison between run and reference data can then be used as
objective evidence of the quality of the modelling.

In commercial terms, the first priority is always the
objective measurement of contract completion relative to the
specification. However, in order to establish the simulators
fitness for purpose, it is also essential that an extensive
validation base should be developed. Without this it would
be difficult to design operator training courses with the
required confidence in simulator response.

Ultimately the simulator should be judged against the
real plant. However, by its very nature, there is little or
no direct time dependent plant data in many areas of major
training interest and it is impossible to formulate acceptance
criteria based on comparison with the real plant. Moreover
due to the myriad permutations of plant conditions, faults,
operator actions, etc., any acceptance and validation tests
must necessarily only form a small subset of the total possibilities.

To mitigate these problems, simulators should be designed
to represent the plant, where possible, according to the fundamental processes involved - be they electrical, electronic,
nuclear or thermohydraulic - and they should draw heavily on
design codes and expertise wherever possible.

VII.   COMPUTER SYSTEM ARCHITECTURE

Evolution of established standards in the computer industry has been rapid and has passed through a number of
identifiable generations. Each of these generations has been
characterised by some common feature, initially derived from
the hardware technology used, but becoming increasingly
difficult to identify as the size and diversity of the industry

has grown. First generation systems could be recognised by the use of thermionic vacuum tubes as the active elements. Second generation designs moved into discrete transistors and integrated circuits heralded the third generation. More abstract architectural features characterise the current range of fourth generation systems and the exact nature of the common features has never been adequately defined. In consequence the whole concept of classifying computer systems by generation went out of fashion until, that is, the Japanese announced their plans for the fifth generation which promises to introduce an impressive range of distinctive features.

Examination of some of the fifth generation concepts puts the first four generations into a common perspective in that they can all be described in terms of "Central Processing Units" (CPUs), "terminals" and "peripherals". These classical terms reflect what could well be described as an "egocentric" approach in that the CPU is often referred to with some reverence as a kind of "egocentre" where the important things happen and the other system components are where processes start out, terminate or merely sit on the periphery waiting for some menial task to be dispensed from the centre.

In the days when a state of the art CPU represented a substantial investmnet and memory prices stood at "a dollar a byte" such views evolved with considerable justification to reflect the relative values of the components. Subsequent developments in telecommunications have reinforced the egocentric view by extending access to expensive facilities to a much wider population of users who get valuable service from devices even further divorced from the egocentre and known affectionately as "remote terminals".

As the Japanese have graphically demonstrated, these classical concepts which have seemed so natural for years do not now reflect any real underlying system or financial constraints. However, they are extremely well developed and supported and it is not impossible that the fifth generation could parallel the attempts to replace the water cooled reciprocating internal combustion engine with a revolutionary rotary one or an electric motor. Once a system has evolved to fulfil a requirement it has an inherent advantage over under developed alternatives which may well outweigh the benefits of a superior underlying principle.

This forthcoming contest and its eventual outcome is of considerable interest to the computing community at large both

as an interesting philosophical puzzle and in terms of
formulating a reasonable model of the future. There is,
however, another faction who cannot now be ignored.

A historical review of the development of computer
science and, in particular, its application to engineering,
reveals a number of repeating patterns. The early machines
of the late 1950s with their delay line memories and large
banks of flashing lights bred a generation of engineers who
wrote programs in binary to assist in the design of anything
from washing machines to nuclear power stations and who
became expert in optimising delay cycles within the machine.
These early systems rapidly evolved into the mainframes and
high level languages replaced the binary codes.

The development of the integrated circuit which spawned
the third generation of mainframes also gave birth to the
minicomputer in the late 1960s and because many of these
machines were too small to support the highly evolved mainframe systems, machine code programming skills were given a
new lease of life. As the technology developed the mini
followed the path pioneered by the mainframe and todays
supermini outperforms yesterdays mainframe by an order of
magnitude.

A third pass through this sequence is now nearing completion as the microprocessor comes of age and can support
high level systems.

It is tempting to speculate that another order of magnitude increase in the power of the home computer is only a
few years away and, therefore, the facilities offered by
todays mainframes at considerable expense will be freely
available to interested individuals on their own personal
equipment.

In this kind of environment any individual or organisation producing software based products must adopt a survival
strategy for the protection of his investment. The commonest
strategy involves the standardisation of the computing
environment at some level in the expectation that upward
compatible hardware will always be available. If this policy
is carried out on a wide scale it establishes the market and
feeds back into the hardware design process.

Given the existence of a fundamental limit to the speed

of a single processor, it is not an unreasonable speculation that this process converges on to a supermicro home computer of the type currently on the horizon.

What then of the fifth generation? Any new machine can usually emulate its predecessor but often the improvement in this mode is marginal. The true benefits come from the non-upward compatible and completely new features. Perhaps it is too early to say what benefit developments in artificial intelligence, speech recognition and the like will bring beyond the obvious application to computer games and word processors.

A. Parallelism

All early digital computers were serial in operation in that the arithmetic units which performed the basic processing operations of addition and subtraction were one bit wide. Integers were represented as sequences of binary digits which were routed through the arithmetic unit which produced results one bit at a time. The processing speed of these machines depended upon the wordlength and the clock rate at which the processing unit operated.

With the introduction of semiconductor logic it became possible to provide parallel arithmetic units which included logic for each of the bit positions in the word and they could therefore perform arithmetic in parallel. Carry look ahead units were also developed which eliminated delays due to carry propagation between bits. These units operated in parallel with the addition/subtraction logic.

Modern microprocessors consisting of a single integrated circuit include logic of this type with wordlength of up to 32 bits and clock speeds of 10 MHz.

Mainframe developments have included a series of further enhancements which, when allied with improvements in device logic, have been responsible for a dramatic increase in performance. Floating point operations which remove the complexity of integer scaling can be performed by co-processors operating in parallel with the integer unit and more than one such unit can be fitted. Pipelines can be formed, again by parallel logic, into which an instruction stream passes with an enhanced throughput because, in effect, the processor is obeying more than one instruction at a time.

As the speed of the central processors increased, larger, highspeed memories were required to provide them with instructions and data at a compatible rate. This led to memory banks with addresses interleaved on low order address bits and capable of parallel operation. Such arrangements provided the processor with data streams from consecutive memory locations at a rate which was higher than the basic memory speed and a function of the interleave factor.

With the advent of high speed semiconductor memories, cache memories were introduced to exploit the statistical nature of memory access. In a typical system, instructions and data extracted from a slow, bulk, ferrite core memory were automatically retained in the semiconductor cache which was equipped with a special addressing scheme suitable for the purpose. A subsequent access of the same data was satisfied from the highspeed memory with an increase in speed which was a function of the repeated access statistics. Updating transfers to bulk memory could also take place in parallel with processor/cache transfers.

Since all of these mechanisms are designed to improve system throughput without altering the functional specification of the computing environment, they are transparent to the software and can be introduced without affecting upward compatibility. Perhaps because of this, none of these architectures are now referred to as parallel.

Since many of the scientific and engineering tasks which require extensive amounts of computation can be formulated in terms of matrix algebra, machines which can operate on data vectors have been developed. These are the parallel machines of the 1980s and they tend to be complex, specialised and expensive. While many of them can present the user with a FORTRAN environment, they are not significantly faster in this mode and their full potential is only realised if correspondingly parallel software is employed.

Apart from a number of notable exceptions, including systems used for some military purposes and weather forecasting, all of these systems are designed as general purpose systems onto which problems are mapped by compilers.

B. Software

It is often the case that dedicated tools provide a more
efficient solution to the problems that they were designed
for than general purpose equivalents. However, the degree
of dedication inevitably limits the range of applications to
which a tool can usefully be applied.

In recognition of these principles, many computer
packages are now designed to permit a degree of adaptability
in the level in which they are programmed. The classic
example of this is the inclusion of machine code blocks in
high level programs. Such blocks, which exploit particular
features of the machines architecture, are therefore dedicated
to it and will not necessarily operate on the next hardware
upgrade. They can however easily be replaced with an
equivalent block for the new machine or coded back to the
high level if the appropriate capacity is available.

Such optimisation processes offend the purist but the
increase in process efficiency can be spectacular and an
order of magnitude improvement is not unusual.

In applying this kind of system design philosophy to
Nuclear Training Simulators, advantage was taken of the closed
nature of the computing environments. Initial development,
entirely in machine code, of the first submarine system in
1965 was subsequently (1971) recoded in a high level language
(SOUL - Systems Orientated Utility Language) implemented
especially for the purpose on both the project computer and
the manufacturers compatible mainframe. The objective of
this language was the elimination of all machine code. This
was achieved by a choice of language level which qualifies
as "high" but only just, since the "accumulator" concept was
retained and precedence processing was not provided. Subsequent transfer to a mini, for which a suitable compiler was
also produced, was a relatively painless process.

Since design authority for the entire software system
rested with the Nuclear Simulation Group, controlled evolution
was possible since the user population was small and operated
in a single building. The often voiced counter arguments
favouring industry standard languages have, over the years,
echoed with a surprisingly wide range of alternative
approaches. These have ranged initially from FORTRAN and
ALGOL through CORAL66 and RTL up to the rarified heights

occupied by ADA and then down again through PASCAL via BCPL
to the current favourite "C". This last language has recently
been described as "SOUL-like". The nuclear industries retention of FORTRAN reflects a similar response characteristic
to the products of the computer software industry.

Success of the submarine and other simulators led to a
research programme which carried these principles into the
hardware design. This resulted in a processor design which
was microprogrammed to present a machine code instruction set
which bore a one-to-one relationship with the SOUL language.

This has two advantages. Firstly, inline macros or
system calls are unnecessary since each language command has
a corresponding single entry object file record which specifies a single machine level instruction. A second advantage
is that this completely removes the need for the inclusion
of machine code blocks yet still retains visibility of the
machine processes should this be required.

Machines with this architecture are used for both the
Hunterston and Torness simulators.

C. Closed Environments

If a software product is well established and has many
users then the environment is open in that the applications
the product is supporting cannot be identified rigourously
by its designer or owner. The functional specification of
such products must therefore be generalised and call for the
use of a formal interface. This also implies that such
products are designed in a climate of uncertainty regarding
the exact nature of the problems it will subsequently solve.
Success of the product encourages its widespread use and
this in turn leads to developmental inertia since any changes
must remain transparent to unseen users who have no particular
interest in them.

Clearly open environments provide many advantages,
particularly to casual users of complex products who find
them available at a cost inversely related to the number of
users. However, many of the software components of a real-
time, full-scope plant simulator are of little practical use
to the computing community at large and, as in the case of
military systems, other factors often prohibit general
availability. Adoption of the conventions and practices of

the open community by simulator designers can therefore accrue many of the problems and limitations of the approach but few of the benefits.

If however, the environment is deliberately "closed" then, provided certain predisposing conditions are met, significant benefits result.

Closure of the environment does not imply long term isolation from the rest of the community, it merely means that the engineering team acquires control of its own product in such a way that it can survive without outside support and can unilaterally make and implement any design decision it judges necessary. Once the simulator design has been brought to fruition, this level of control can be passed on to the end user or retained and exercised on his behalf as required. Updates to standard utilities and new packages can be absorbed at the system designers discretion, who can ignore them if he so chooses. He is never in the position of finding that a product on which he is depending is no longer supported or that the current issue violates some of his unique constraints. He can also insist that any modification he requires is implemented in the manner he specifies.

Even though the Hunterston and Torness simulators use compatible computers and many common software modules, both projects are operated as closed environments. This does not prohibit mutually beneficial interchanges but it does enforce strict configuration control.

D. Eminence

Systems designed for use in open environments are usually generalised and there is always motivation to widen the field of application. This encourages the view that this aspect of system design is fundamental and part of an industry wide standard.

If, as in the case of a plant specific simulator, there is little or no general market for identical copies of the design, then there is no substance in this viewpoint. Engineers operating in a closed environment should therefore be directed to solve the problems they actually have rather than ones they might encounter in the future or that they can see arising in other areas.

It is tempting to use the word applications in place of areas in the previous sentence but this, like the egocentric view, is "open" terminology in that it implies that computer technology is the egocentre and is "applied" to problems. The closed view would say that the task to be performed is the egocentre and computer technology may figure in its execution. In particular, it often turns out that parts of the process are not suitable for solution by computer. Under these cir-cumstances the closed system designer should get on and solve his specific problem and not try to bend it into a form suitable for the "application" of the computer.

This is not to say that outside the context of a given project computer-based solutions to such problems should not be implemented. Many of the tasks involved in the production of a practical simulator are processes which could be automated. However, if the effort involved in the development of the automatic system is an order of magnitude greater than that required for a single instance of the solution by manual means, then it is only worth automating if ten such instances are to hand. Designers of open systems have a vested interest in taking the sometimes optimistic view that this is indeed the case.

Another important area to which this type of argument applies relates to the question of optimisation which should be applied to the total process rather than compartmentalised. Many aspects of a complex system can be optimised and these range from the obvious practicalities of cost and timescale, through system performance and reliability, to the job satis-faction of the people involved. These aspects interact in a complex fashion and there is not usually a unique solution which provides all round satisfaction.

Application of the closed systems philosophy to the question of optimisation, benefits from the use of the concept of *Eminence* which serves to quantify (in non-dimensional terms) the significance of various processes as they appear in speci-fic environments.

If a process is, in some way, activated at a high rate then its eminence is also high. Conversely a low rate of activation infers a correspondingly low eminence. In a practical system considerable code is often required to initialise models, data structures and configuration tables. Such code has a low eminence (but not low importance) since a single pass per simulation session is sufficient. High level formulations of such code are therefore appropriate to maximise clarity and ease of production.

At the other extreme any large plant simulation involves the solution of large sets of stiff differential equations and this involves many classical problems. Relatively small sections of computer code often lie at the heart of the implementation which, because of frequent iteration rates and the real time constraint, exhibit a very high Eminence.

Such elements, because of their small size, do not present a significant programming problem but the efficiency of their implementation is crucial.

In the Hunterston and Torness designs the microprocessors all contain microcoded implementations of a series of such Eminent processes which effectively extend the machine level instruction set. As a result, in these areas, the processors act more as dedicated hardware solutions than general purpose computers often with an increase in speed of up to an order of magnitude.

The nature of the Eminent processes is also recognised at a much higher level in the system architecture. Since the environment is closed, the interconnection of processors to suit mathematical characteristics of the plant model can be carried out by designers without violating any hidden constraints. This results in an architecture which is specific for a particular set of Eminent processes but can be assembled from sets of identical standard components. Its specialised nature has, in consequence, little impact on the manufacturing process. Since the simulators are specific to power generation plant on which training exercises take several hours and fill up the timetable, the efficiency with which other, unrelated problems can be solved is a matter of little significance.

VIII. CONCLUSION

A main driving force in the development of computer systems is the rate of change of device technology. This is graphically illustrated by the history of bulk memory technology as depicted in Table III which shows the characteristics of typical single devices over the past decade.

This general pattern of improvement is apparent in all types of microelectronic device and it shows little sign of abating.

Computer science has now come of age and powerful computer systems can be found in many homes. Much of the mystique which surrounded the subject when it first appeared has now evaporated and computing subjects form part of modern High School curricula.

The basic principles of compiler, operating system and general purpose utility design are well established and freely available to the closed system designer.

Many years in the field of computer based system design have resulted in the formulation of a design philosophy which encompasses the total process.

This opens the way for the continuing developments in microelectronics to be applied to large real-time systems by the use of large numbers of small, simple, yet powerful computing elements operating in, subject to an agreed meaning of the word, parallel.

TABLE III

THE DEVELOPMENT OF BULK MEMORY DEVICES

Year	Capacity (k)	Organisation	Cycle time (ns)
1975	1	1k x 1 bit	2000
1978	4	4k x 1 bit	1200
1980	16	16k x 1 bit	650
1982	64	16k x 1 bit	450
1984	256	64k x 4 bits	200
1985	1024	128k x 8 bits	150

ACKNOWLEDGEMENT

The author acknowledges the assistance provided by the CEGB's Systems Engineering Branch in connection with reactor and boiler modelling, and thank management of the SSEB and the Marconi Company for their agreement to this article being published.

Allnoor Allidina studied electrical and electronic engineering at the University of Manchester, Institute of Science and Technology (1977) and subsequently stayed there to complete MSc and PhD studies; he has been a Lecturer at UMIST since 1979. His research interests include Self-Tuning Control, Fault Detection, Parallel Processing for Simulators and the scheduling of large systems in the power and manufacturing industries. Dr Allidina's industrial contacts are presently with Shell, Total, Unilever and the UKAEA.

Geoff Budd has worked for the CEGB since 1963 after graduating from Electronic Engineering at the University of Southampton. For the last eighteen years he has worked at power plant modelling in the CEGB London Headquarters Computing Centre. He now heads the Simulator and Mini/Micro Systems Section, and is responsible for the development of new simulators and the long-term support and enhancement of existing simulators. Budd's real-time mathematical modelling techniques are the foundation for the simulation strategy of the CEGB's four AGR simulators and a dual-purpose fossil plant simulator.

Trevor Chambers studied mathematics at Brunel University followed by work on the simulation of guided weapon systems for EMI Electronics. He is now a Section Head at the CEGB's Computing Centre where his team provides comprehensive support for the digital simulators serving the Board's engineering design needs.

Barry Daniels is a Project Manager at the National Centre of Systems Reliability operated by the United Kingdom Atomic Energy Authority. He has 20 years experience in the application and assessment of computers in industry. His interests include the use of simulation to study performance and reliability for a wide range of engineering systems. He is Vice-Chairman of the European Workshop on Industrial Computer Systems Committee on Safety, Security and Reliability of Computers.

219

Max Jervis graduated from the University of Manchester to work on general instrumentation development at the General Electric Company UK, Taylor Tunnicliff and the research laboratories of Associated Electrical Industries. His entry into nuclear power was via the AEI-John Thompson Consortium that subsequently went to make up the National Nuclear Corporation where he worked on control and instrumentation for Magnox reactors. Jervis joined the CEGB in 1964 and is now C and I Projects and Systems Engineer in the CEGB Generation Development and Construction Division. He previously contributed to Volume 11 of the series as well as to the Institution of Electrical Engineers.

Professor Madan G Singh is a graduate of the University of Exeter, followed by a Cambridge PhD (1973) where he was a Fellow of St John's. His research interests in complex systems have found application in both engineering systems and management systems. Professor Singh has been at UMIST from 1979 but also has held visiting appointments at INSEAD, Fontainebleau. He is editor-in-chief of the Encyclopaedia of Systems and Control.

Robert B Stammers graduated with a BSc and PhD in Psychology from the University of Hull. He is now a Lecturer at Aston University and a Fellow of the Ergonomics Society. Early research centred on psychological aspects of industrial training and he now combines these interests with the study of computing technology in training systems. These studies led to his well known textbook on the psychology of training.

John Wiltshire is currently Chief Systems Engineer of Marconi Instrument's Nuclear Power Simulation Group. He was originally a mathematician (University of London) and first worked for Plessey in telecommunications leading to simulation of modern telephone switching systems. In 1965 John joined Elliott Brothers to work on Royal Navy nuclear submarine training simulators, an appointment leading in turn to Technical Project Manager of the Hunterston 'B' AGR Simulator.

Mike Whitmarsh-Everiss' early career was in the Plastics Industry. He obtained workshop, drawing office and design office experience with Williams and James, manufacturers of hydraulic, vacuum and pneumatic equipment. A period working on radar for the Royal Air Force was followed by employment with Rotol Ltd in the design of variable pitch propellors, turbo-alternators and ducted windmills. In 1958 Mike joined the Nuclear Plant Design Branch of the CEGB. He is currently Head of the Plant Kinetics Group at Barnwood. Whitmarsh-Everiss' work in this time has been directed to the control, operational and fault analysis of a wide variety of nuclear plant and he has recently accepted similar responsibilities for conventional plant.